SpringerBriefs in Computer Science

SpringerBriefs present concise summaries of cutting-edge research and practical applications across a wide spectrum of fields. Featuring compact volumes of 50 to 125 pages, the series covers a range of content from professional to academic.

Typical topics might include:

- A timely report of state-of-the art analytical techniques
- A bridge between new research results, as published in journal articles, and a contextual literature review
- A snapshot of a hot or emerging topic
- An in-depth case study or clinical example
- A presentation of core concepts that students must understand in order to make independent contributions

Briefs allow authors to present their ideas and readers to absorb them with minimal time investment. Briefs will be published as part of Springer's eBook collection, with millions of users worldwide. In addition, Briefs will be available for individual print and electronic purchase. Briefs are characterized by fast, global electronic dissemination, standard publishing contracts, easy-to-use manuscript preparation and formatting guidelines, and expedited production schedules. We aim for publication 8–12 weeks after acceptance. Both solicited and unsolicited manuscripts are considered for publication in this series.

More information about this series at http://www.springer.com/series/10028

Shenglin Zhao · Michael R. Lyu
Irwin King

Point-of-Interest Recommendation in Location-Based Social Networks

 Springer

Shenglin Zhao
Youtu Lab
Tencent
Shenzhen, Guangdong, China

Michael R. Lyu
Department of Computer
  Science and Engineering
The Chinese University
  of Hong Kong
Hong Kong, China

Irwin King
Department of Computer
  Science and Engineering
The Chinese University
  of Hong Kong
Hong Kong, China

ISSN 2191-5768   ISSN 2191-5776   (electronic)
SpringerBriefs in Computer Science
ISBN 978-981-13-1348-6   ISBN 978-981-13-1349-3   (eBook)
https://doi.org/10.1007/978-981-13-1349-3

Library of Congress Control Number: 2018947499

Printed on acid-free paper

This Springer imprint is published by the registered company Springer Nature Singapore Pte Ltd.
The registered company address is: 152 Beach Road, #21-01/04 Gateway East, Singapore 189721,
Singapore

Preface

Location-based social networks (LBSNs) have become popular recently because of the explosive increase of smart phones that makes users easily to access to the LBSN Apps. More than 2.3 billion people worldwide use smartphones in 2017 as predicted by EMarketer, which prospers the online LBSNs. A typical LBSN such as Foursquare collects users' check-in information including visited locations' geographical information (latitude and longitude) and users' comments at the location and allows users to make friends and share information as well. Driven by the collected big data in LBSNs, point-of-interest (POI) recommendation arises to improve the user experience in the App, which attempts to suggest each user a list of POIs that the user may feel interesting and be willing to visit in the future.

Developing POI recommendation systems requires analytics of the human mobility with respect to real-world POIs. Different from watching on Netflix or shopping on Amazon, checking-in at a POI in LBSNs is a physical activity, which causes the most important feature in POI recommendation: geographical influence. In addition, check-ins exhibit specific temporal characteristics. For instance, users check-in at POIs around the office in the daytime while at bars in the evening. These geographical and temporal features make the POI recommendation more challenging than traditional recommendation systems.

In this book, we systematically study the problem of POI recommendation in LBSNs. In particular, we analyze the user mobility in LBSNs from geographical and temporal perspectives and further develop POI recommendation systems. First, we analyze the user mobility in LBSNs from geographical and temporal perspective, respectively, and show how to capture the geographical and temporal influence in a POI recommendation system. Then, we develop two POI recommendation systems: Geo-Teaser and STELLAR. Finally, we conclude this book and point out future work directions.

This book is intended for professionals involved in POI recommendation and graduate students working on the location-based services-related problems. It is assumed that the reader has a basic knowledge of mathematics, as well as a certain background in recommendation systems. The reader can get an overview of the POI recommendation research area. We hope this monograph will be a useful reference for students, researchers, and professionals to understand basic methodologies of POI recommendations in LBSNs. This book can be used as a starting point for POI recommendation research topics.

Hong Kong, China   Shenglin Zhao
February 2018   Michael R. Lyu
                                                                       Irwin King

Contents

Chapter 1
Introduction

Abstract This chapter provides an overview of POI recommendation in LBSNs, including backgrounds, related work, and organizations of this book.

Keywords POI recommendation · Spatio-temporal data · Location-based services

1.1 Overview

Location-based services play an important role in this Internet of Things (IoT) era. To monitor the status of devices connected to the Internet, analyze the collected data from different kinds of devices, and provide personalized services for device users, the location information of the device is indispensable for data analysis. For instance, the smart watch collects the location information of users and records their daily trajectories; and the map applications such as Google Maps in smart phones collect users' location information and guide users to anywhere in real-time. Figure 1.1 shows the location-based services in six aspects: search and recommendation, transportation, healthcare, public safety, game, environment monitoring, etc. Specifically, the location-based search and recommendation work for two kinds of applications: search engines such as Google and Baidu and LBSNs such as Yelp and Foursquare. In this book, we focus on the search and recommendation task in LBSNs.

LBSNs such as Foursquare and Facebook Places allow users to share their check-in behaviors, make friends, and write comments on visited locations, also called POIs [5, 56]. LBSNs are very popular now—for instance, Foursquare has attracted over 55 million people worldwide to use its service each month and recorded over 12 billion check-ins in total until 2018.[1] To improve user experience in LBSNs by suggesting favorite locations, a typical search and recommendation task namely POI recommendation [4, 9, 57, 61, 75, 76] comes out, which mines users' check-in sequences to recommend places where an individual may feel interested and be willing to check-in in the future. The POI recommendation applications are of

---

[1] https://foursquare.com/about.

S. Zhao et al., *Point-of-Interest Recommendation in Location-Based Social Networks*, SpringerBriefs in Computer Science, https://doi.org/10.1007/978-981-13-1349-3_1

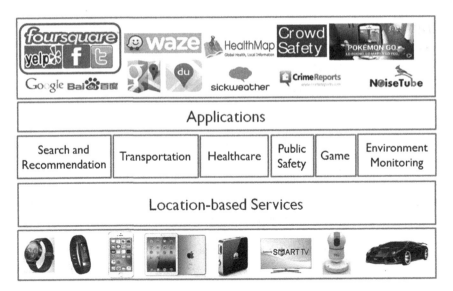

Fig. 1.1 Demonstration of location-based services

significance in two aspects: helping users explore new interesting places in a city and facilitating business owners to launch advertisements to the target customers.

Developing POI recommendation systems requires analyzing the human check-in activity in LBSNs. The check-in activity represents the user interactions with real-world POIs and exhibits specific geographical and temporal characteristics. From the geographical perspective, most of the check-ins happen in some constrained regions such as the district around the user's home or office. From the temporal perspective, the check-in activity also exhibits some specific patterns. For instance, users check-in at POIs around the office in the day time while bars in the evening. These unique features make the POI recommendation different from traditional recommendation systems. Hence, we need to comprehensively understand the human mobility in LBSNs and develop new algorithms for POI recommendation.

In this book, we study the POI recommendation in LBSNs. In particular, we review the literature in POI recommendation, analyze the user mobility in LBSNs, and develop POI recommendation systems. First, we show the overview and background of POI recommendation problems in LBSNs. Second, we analyze the user mobility in LBSNs from geographical and temporal perspective respectively and show how to capture the geographical and temporal influence to enhance the POI recommendation system. Third, we develop a **Geo-te**mporal sequential **e**mbedding **r**ank (Geo-Teaser) model for POI Recommendation. Fourth, we develop a spatial-**te**mporal **la**tent **r**anking (STELLAR) model for successive POI recommendation. Finally, we conclude this book and point out possible future work.

1.2 Backgrounds

In this section, we first show the problem description of POI recommendation. Then, we analyze the user check-in behavior in LBSNs and show how to take advantage of the behavior characteristics to model the POI recommendation task.

1.2.1 Problem Description

POI recommendation aims to mine users' check-in records and recommend POIs for users in LBSNs. Formally, we define two important terms, i.e., check-in and check-in sequence, as follows.

Definition 1.1 (*Check-in*) A check-in is denoted as a triple $\langle u, l, t \rangle$ that depicts a user u visiting POI l at time t.

Definition 1.2 (*Check-in sequence*) A check-in sequence is a set of check-ins of user u, denoted as $S_u = \{\langle l_1, t_1 \rangle, \cdots, \langle l_n, t_n \rangle\}$, where t_i is the check-in timestamp. For simplicity, we denote $S_u = \{l_1, \cdots, l_n\}$.

POI recommendation recommends a user a list of POIs via mining the check-in records. Given Definitions 1.1 and 1.2, the problem of POI recommendation can be defined as follows.

Definition 1.3 (*POI recommendation*) Given all users' check-in sequences S, POI recommendation aims to recommend a POI list S_N to each user u. Here S is a collected check-in sequence set, contain all sequences S_u for all users.

POI recommendation is a branch of recommendation systems, which encourages to address this task through borrowing ideas from conventional recommendation systems such as movie recommendation. Hence, conventional recommendation system techniques are used to recommend POIs, e.g., collaborative filtering (CF) methods. However, the special scenario in LBSNs that the location bridges the physical world and the online networking services, arouses new challenges to the traditional recommendation system techniques. Take Foursquare as an example, Fig. 1.2 demonstrates how the check-in information is recorded, including user name, POI, check-in timestamp, and geographical information in the map. After introducing the location, some new challenges appear, which can be summarized as follows [72].

1. Physical constraints. The check-in activity is limited by physical constraints, compared with shopping online from Amazon and watching a movie on Netflix. For one thing, users in LBSNs check-in at geographically constrained areas. As observed in [5, 6], users usually check-in at POIs around their homes and offices, and there are a few check-ins out of their cities. For another, shops regularly provide services at some limited time. For instance, most of coffee shops open

Fig. 1.2 Demonstration of
check-in information in
Foursquare

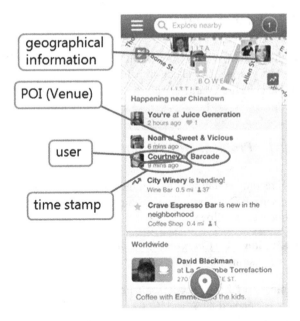

during day time but close at night. Such physical constraints make the check-in activity in LBSNs exhibit significantly spatial and temporal properties [1, 5, 12, 13, 39, 55, 60, 73, 74].

2. Extreme sparseness. A typical user in LBSNs such as Foursquare contains hundreds of check-ins each year. But there are millions of POIs in the LBSNs. Compared with traditional movie recommendation, POI check-in data are much sparser. So it is difficult to suggest top N (usually five or ten) POIs from millions of candidates.

3. Complex relations. The location sharing activities in the online social media alter original social relations since people are apt to make new friends with geographical neighbors [45, 46]. Moreover, for online social media services such as Twitter and Facebook, the location for geo-tagging yields new relations between locations and locations [63], and as well between users and locations [11, 44, 58].

4. Heterogeneous information. LNSNs consist of different kinds of information, including not only check-in records, the geographical information of locations, and venue descriptions but also users' social relation information and media information (e.g., user comments and tweets). The heterogeneous information depicts the user activity from a variety of perspectives [52, 53, 66], inspiring POI recommendation systems of different kinds [26, 27, 30, 35, 43, 51, 65].

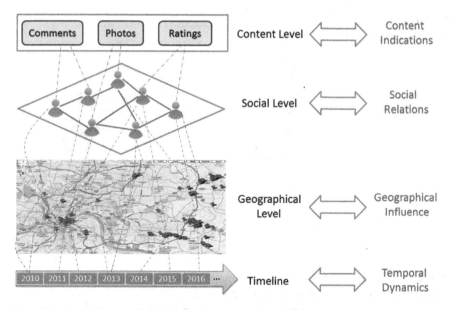

Fig. 1.3 Influential factors in LBSNs

1.2.2 User Behavior Analysis

Analyzing the user check-in activity implies to find influential factors of POI recommendation. Because of the spatial and temporal properties resulted from the physical constraints and heterogeneous information such as locations' geographical information and users' comments, the check-in activity is a synthesized decision from a variety of factors. Figure 1.3 shows four main factors in POI recommendations: temporal dynamics, geographical influence, social relations, and content indications. This section will demonstrate how each factor affects the check-in activity and how to model each influential factor for POI recommendation.

1.2.2.1 Geographical Influence

Geographical influence is an important factor that distinguishes the POI recommendation from traditional item recommendation because the check-in behavior depends on locations' geographical features [3, 24, 30, 57]. Analysis on users' check-in data shows that a user acts in geographically constrained areas and prefers to visiting POIs nearby where the user has checked-in. The feature of geographical constraints can shrink the POI candidate set and alleviate the effect of data sparsity [3, 68, 71, 74, 76]. Several studies [3, 8, 24, 30, 50, 57, 62, 64, 67, 68, 71, 74, 76] attempt to employ the geographical influence to improve POI recommendation systems. In particular, three representative models, i.e., power law distribution model, Gaussian

Fig. 1.4 Power law
distribution pattern [57]

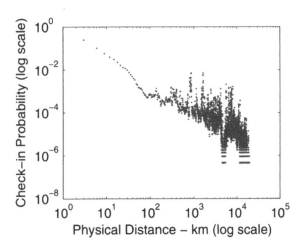

distribution model, and kernel density estimation model, are proposed to capture the geographical influence in POI recommendation.

In [57], Ye et al. employ a power law distribution model to capture the geographical influence. Power law distribution pattern has been observed in human mobility such as withdraw activities in ATMs and travel in different cities [2, 14, 39]. Also, Ye et al. discover a similar pattern of users' check-in activity in LBSNs [56, 57]. Figure 1.4 demonstrates two POIs' co-occurrence probability distribution over the distance between two POIs. Because of the power law distribution in Fig. 1.4, we are able to model the geographical influence as follows. The co-occurrence probability y of two POIs by the same user can be formulated as follows,

$$y = a * x^b, \tag{1.1}$$

where x denotes the distance between two POIs, a and b are parameters of the power-law distribution. Here, a and b should be learned from the observed check-in data, depicting the geographical feature of the check-in activity. A standard way to learn the parameters, a and b, is to transform Eq. (1.1) to a linear equation via a logarithmic operation, and learn the parameters by fitting a linear regression problem.

On the basis of the geographical influence model depicted through the power law distribution, new POIs can be suggested according to the following formula. Given a checked-in POI set L_i, the probability $Pr(l_j|L_i)$ of visiting POI l_j for user u_i, is formulated as,

$$Pr(l_j|L_i) = \frac{Pr(l_j \cup L_i)}{Pr(L_i)} = \prod_{l_y \in L_i} Pr(d(l_j, l_y)), \tag{1.2}$$

where $d(l_j, l_y)$ denotes the distance between POI l_j and l_y, and $Pr(d(l_j, l_y)) = a * d(l_j, l_y)^b$. In [56, 57], Ye et al. leverage the power law distribution to model the

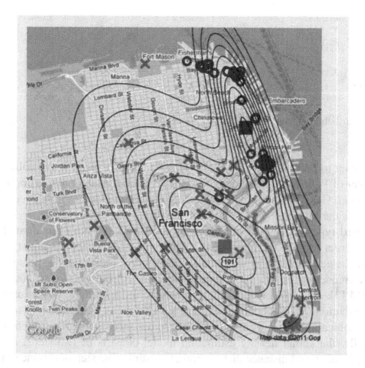

Fig. 1.5 Check-in distribution in multi-centers [6]

geographical influence and combine it with collaborative filtering techniques [40] to recommend POIs. In addition, Yuan et al. [64] also adopt the power law distribution model, but learn the parameter using a Bayesian rule instead.

The second type to model the geographical influence is a series of Gaussian distribution based methods. Cho et al. [6] observe that users in LBSNs always act around some activity centers, e.g., home and office, as shown in Fig. 1.5. Further, Cheng et al. [3] propose a Multi-center Gaussian Model (MGM) to capture the geographical influence for POI recommendation. Given the multi-center set C_u, the probability of visiting POI l by user u is defined by

$$P(l|C_u) = \sum_{c_u=1}^{|C_u|} P(l \in c_u) \frac{f_{c_u}^\alpha}{\sum_{i \in C_u} f_i^\alpha} \frac{N(l|\mu_{C_u}, \sum_{C_u})}{\sum_{i \in C_u} N(l|\mu_i, \sum_i)}, \tag{1.3}$$

where $P(l \in c_u) \propto \frac{1}{d(l,c_u)}$ is the probability of the POI l belonging to the center c_u, $\frac{f_{c_u}^\alpha}{\sum_{i \in C_u} f_i^\alpha}$ denotes the normalized effect of the check-in frequency on the center c_u and parameter α maintains the frequency aversion property, $N(l|\mu_{C_u}, \sum_{C_u})$ is the probability density function of Gaussian distribution with mean μ_{C_u} and covariance matrix $\sum_{C_u}$. Specifically, the MGM employs a greedy clustering algorithm on the

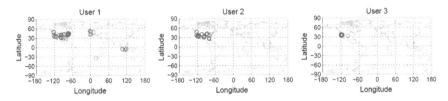

Fig. 1.6 Distributions of personal check-in locations [67]

check-in data to find the user activity centers, which may result in the unbalanced assignment of POIs to different activity centers. Hence, Zhao et al. [71] propose a genetic-based Gaussian mixture model to capture the geographical influence, which outperforms the MGM in POI recommendation.

The third type of geographical model is the kernel density estimation (KDE) model. In order to mine the personalized geographical influence, Zhang et al. [67] argue that the geographical influence on each individual user should be personalized rather than modeling through a common distribution, e.g., power law distribution [57] and MGM [3]. As shown in Fig. 1.6, it is hard to model different users using the same distribution. To this end, they leverage kernel density estimation [47] to model the geographical influence using a personalized distance distribution for each user. Specifically, the kernel density estimation model consists of two steps: distance sample collection and distance distribution estimation. The step of distance sample collection generates a sample X_u for a user by computing the distance between every pair of locations visited by the user. Then, the distance distribution can be estimated through the probability density function f over distance d,

$$f(d) = \frac{1}{|X_u|\sigma} \sum_{d' \in X_u} K(\frac{d - d'}{\sigma}),$$ (1.4)

where σ is a smoothing parameter, called the bandwidth. $K(\cdot)$ is the Gaussian kernel

$$K(x) = \frac{1}{\sqrt{2\pi}}e^{-\frac{x^2}{2}}.$$ (1.5)

Denote $L_u = \{l_1, l_2, \ldots, l_n\}$ as the visited locations of user u. The probability of user u visiting a new POI l_j given the checked-in POI set L_u is defined as,

$$p(l_j|L_u) = \frac{1}{|L_u|} \sum_{l_i \in L_u} f(d_{ij}),$$ (1.6)

where d_{ij} is the distance between l_i and l_j, $f(\cdot)$ is the distance distribution function in Eq. (1.4).

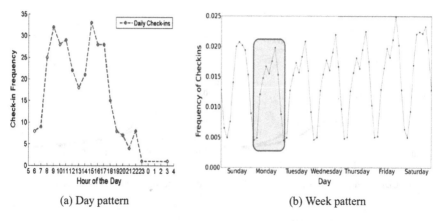

(a) Day pattern   (b) Week pattern

Fig. 1.7   Periodic pattern [5]

1.2.2.2   Temporal Influence

Temporal influence is of vital importance for POI recommendation because physical constraints on the check-in activity result in specific patterns. Temporal influence in a POI recommendation system performs in three aspects: periodicity, consecutiveness, and non-uniformness.

Users' check-in behaviors in LBSNs exhibit the periodic pattern. For instance, users always visit restaurants at noon and have fun in nightclubs at night. Also, users visit places around the office on weekdays and spend time in shopping malls on weekends. Figure 1.7 shows the periodic pattern in a day and a week, respectively. The check-in activity exhibits this kind periodic pattern—visiting the same or similar POIs at the same time slot. This observation inspires the studies exploiting this periodic pattern for POI recommendation [6, 9, 64, 69].

Consecutiveness performs in the check-in sequences, especially in the successive check-ins. Successive check-ins are usually correlated. For instance, users may have fun in a nightclub after dining in a restaurant. This frequent check-in pattern implies that the nightclub and the restaurant are geographically adjacent and correlated from the perspective of venue function. Data analysis on Foursquare and Gowalla in [77] explores the spatial and temporal property of successive check-ins in Fig. 1.8, namely the complementary cumulative distributive function (CCDF) of intervals and distances between successive check-ins. It is observed that many successive check-ins are highly correlated: over 40 and 60% successive check-in behaviors happen in less than 4 h in Foursquare and Gowalla respectively; about 90% successive check-ins happen in less than 32 km (half an hour driving distance) in Foursquare and Gowalla. Researchers exploit the Markov chain model to capture the sequential pattern [4, 7, 16, 70]. Studies in [4, 7] assume that two successive checked-in POIs in a short term are highly correlated and employ the factorized personalized Markov chain (FPMC) model [38] to recommend successive POIs. Zhang et al. [70] propose an additive

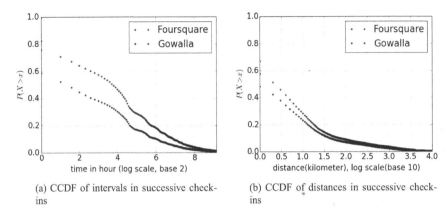

(a) CCDF of intervals in successive check-ins

(b) CCDF of distances in successive check-ins

Fig. 1.8 Consecutive pattern [77]

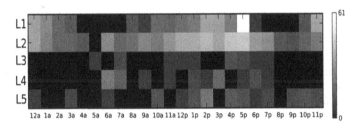

Fig. 1.9 Demonstration of non-uniformness [12]

Markov model to learn the transitive probability between two successive check-ins. Zhao et al. [77] exploit a spatial temporal latent ranking model for POI recommendation, which captures the consecutiveness by a POI-POI latent interaction similar to FPMC model.

The non-uniformness feature depicts a user's check-in preference variance at different hours of a day, or at different months of a year, or at different days of a week [9]. As shown in Fig. 1.9, the study in [9] demonstrates an example of a random user's aggregated check-in activities on the user's top five most visited POIs. It is observed that a user's check-in preference changes at different hours of a day—the most frequent checked-in POI alters at different hours. Similar temporal characteristics also appear at different months of a year, and different days of a week as well. This non-uniformness feature can be explained from the user's daily life customs: (1) A user may check-in at POIs around the user's home in the morning hours, visit places around the office in the day hours, and have fun in bars during night hours. (2) A user may visit more locations around the user's home or office on weekdays. On weekends, the user may check-in more at shopping malls or vacation places. (3) At different months, a user may have different hobbies for food and entertainment. For instance, a user would visit ice cream shops in the months of summer while visit hot pot restaurants in the months of winter.

1.2.2.3 Social Influence

Inspired by the assumption that friends in LBSNs share more common interests than non-friends, social influence is explored to enhance POI recommendation [3, 11, 12, 22, 54, 56, 68, 70]. In fact, employing social influence to enhance recommendation systems has been explored in traditional recommendation systems, both in memory-based methods [19, 34] and model-based methods [20, 32, 33]. Researchers borrow the ideas from traditional recommendation systems to POI recommendation. In the following, we demonstrate representative studies capturing social influence in two aspects: memory-based and model-based.

Ye et al. [56] propose a memory-based model, friend-based collaborative filtering (FCF) approach for POI recommendation. FCF model constrains the user-based collaborative filtering to find top similar users in friends rather than all users of LBSNs. Hence, the preference r_{ij} of user u_i at l_j is calculated as follows,

$$r_{ij} = \frac{\sum_{u_k \in F_i} r_{kj} w_{ik}}{\sum_{u_k \in F_i} w_{ik}}, \tag{1.7}$$

where F_i is the set of friends with top-n similarity, w_{ik} is similarity weight between u_i and u_k. FCF enhances the efficiency by reducing the computation cost of finding top similar users. However, it overlooks the non-friends who share many common check-ins with the target user. Experimental results show that FCF brings very limited improvements over user-based POI recommendation in terms of precision.

Cheng et al. [3] apply the probabilistic matrix factorization with social regularization (PMFSR) [33] in POI recommendation, which integrates social influence into PMF [42]. Denote $\mathcal{U}$ and $\mathcal{L}$ are the set of users and POIs, respectively. PMFSR learns the latent features of users and POIs by minimizing the following objective function,

$$\underset{U,L}{\arg\min} \sum_{i=1}^{|\mathcal{U}|} \sum_{j=1}^{|\mathcal{L}|} I_{ij}(g(c_{ij}) - g(U_i^T L_j))^2 + \lambda_1 ||U||_F^2 + \lambda_2 ||L||_F^2 +$$
$$\beta \sum_{i=1}^{|\mathcal{U}|} \sum_{u_f \in F_i} sim(i,f) ||U_i - U_f||_F^2, \tag{1.8}$$

where c_{ij} is the check-in frequency, U_i, U_f, and L_j are the latent features of user u_i, u_f, and POI l_j respectively, I_{ij} is an indicator denoting user u_i has checked-in at POI l_j, F_i is the set of user u_i's friends, $sim(i, f)$ denotes the social weight between user u_i and u_f, and $g(\cdot)$ is the sigmoid function to mapping the target value into the range of [0,1]. In this framework, the social influence is incorporated by the social constraints that ensure latent features of friends keep in close at the latent subspace. Due to its validity, Yang et al. [54] also employ the same framework to their sentiment-aware POI recommendation.

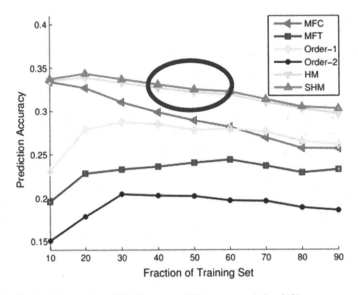

Fig. 1.10 The significance of social influence on POI recommendation [12]

Although social influence improves traditional recommendation system significantly [20, 32, 33], the social influence on POI recommendation shows limited
improvements [3, 12, 56]. Figure 1.10 shows the limited improvement achieved
from social influence in [12]. Why this happens can be explained as follows. Users
in LBSNs make friends online without any limitation; on the contrary, the check-in
activity requires physical interactions between users and POIs. Hence, friends in
LBSNs may share the common interest but may not visit common locations. For
instance, friends in favor of Italian food from different cities will visit their own
local Italian food restaurants. This phenomenon differs from the online movie and
music recommendation scenarios in Netflix and Spotify.

1.2.2.4   Content Indications

In LBSNs, users generate contents including tips and photos about the POIs. Although
contents do not accompany each check-in record, the available contents such as the
user comments and photos, can be used to enhance the POI recommendation [10,
17, 23, 49, 54, 59]. On the one hand, user comments provide extra information from
the shared tips beyond the check-in behavior, e.g., the preference on a location. For
instance, the check-in at an Italian restaurant does not necessarily mean the user
likes this restaurant. Probably the user just likes Italian food but not this restaurant,
even dislikes the taste of this restaurant. Compared with the check-in activity, the
comments usually provide explicit preference information, which is a kind of complementary explanations for the check-in behavior. As a result, the comments are

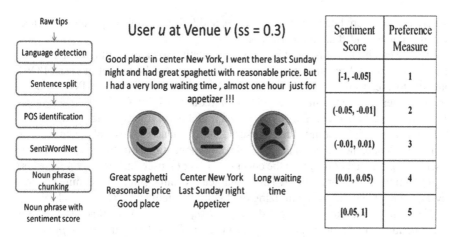

Fig. 1.11 Sentiment-preference transforming rule [54]

able to be used to deeply understand the users' check-in behavior and improve POI recommendation [10, 17, 54]. On the other hand, photos about POIs also reveal users' check-in preference. For example, a user who posts many architecture photos is more likely to visit famous landmarks; while a user posts lots of images about food has more incentive to visit restaurants. Thus, images have potentials to improve the performance of POI recommendation. In the following, we report two representative studies that exploit comments and photos to enhance the POI recommendation respectively.

The research in [54] is the first and representative work exploiting the comments to strengthen the POI recommendation. Yang et al. [54] propose a sentiment-enhanced location recommendation method, which utilizes the user comments to adjust the check-in preference estimation. As shown in Fig. 1.11, the raw tips in LBSNs are collected and analyzed using natural language processing techniques, including language detection, sentence split, POS identification, processed by SentiWordNet, and Noun phrase chunking. Then, each comment is given a sentiment score. According to the estimated sentiment, a preference score of one user at a POI is generated. Figure 1.11 also shows how to handle a comment example: transforming it to several noun phrases such as "Reasonable price", "Good place", and "Long waiting time", generating a sentiment score of 0.3, and mapping this value to the preference measure of 5. Moreover, through combining the preference measure from sentiment analysis and the check-in frequency, the proposed model in [54] generates a modified rating $\hat{C}_{i,j}$ measuring the preference of user u_i at a POI l_j. Accordingly, the traditional matrix factorization method can be employed to recommend POIs through the following objective,

$$\arg\min_{U,L} \sum_{(i,j)\in\Omega} (\hat{C}_{i,j} - U_i L_j^T)^2 + \alpha||U||_F^2 + \beta||L||_F^2, \tag{1.9}$$

where U_i and L_j are latent features of user u_i and l_j respectively, $\hat{C}_{i,j}$ is the combined rating value, α and β are regularizations.

The research in [49] is the first and representative work exploiting the photos to strengthen the POI recommendation. Let $\mathcal{U}$, $\mathcal{L}$, and $\mathcal{P}$ be the set of users, POIs, and photos, respectively. Furthermore, $\mathbf{X} \in R^{|\mathcal{U}| \times |\mathcal{L}|}$ denotes the user-POI check-in matrix, where each entry means the check-in frequency. Next, $\mathbf{G} \in R^{|\mathcal{U}| \times |\mathcal{L}|}$ denotes the normalized version of $\mathbf{X}$ with $\mathbf{G}_{ij} = g(\mathbf{X}_{ij})$ and $g(\cdot)$ is the sigmoid function. $\mathcal{P}_{u_i}$ denotes the set of photos uploaded by user u_i and $\mathcal{P}_{l_j}$ denotes the set of phones tagged to POI l_j. Hence, the image enhanced POI recommendation aims to recommend each user top k unvisited POIs, given the check-in matrix $\mathbf{G}$, user images $\mathcal{P}_u$ for all users, and POI images $\mathcal{P}_l$ for all POIs. This visual feature enhanced POI recommendation system can be learned by maximizing the likelihood defined on $\mathbf{G}$, $\mathcal{P}_u$, and $\mathcal{P}_l$.

To learn the maximum likelihood on $\mathcal{P}_u$, $P(f_{is} = 1|u_i, p_s)$ is defined to measure the probability that a photo p_s belongs to a user u_i. Considering the image p_s posted by u_i, it is natural to assume that p_s contains certain visual contents that meet u_i's preferences; while for an arbitrary image p_w posted by other users, i.e., $p_w \notin \mathcal{P}_{u_i}$, p_w is less likely to contain visual contents that meet u_i's preferences. Meantime, u_i's preferences are captured by the latent feature $\mathbf{u}_i \in R^K$. Then, the probability $P(f_{is} = 1|u_i, p_s)$ is defined via a softmax function,

$$P(f_{is} = 1|u_i, p_s) = \frac{\exp(\mathbf{u}_i^T \cdot \mathbf{P} \cdot CNN(p_s))}{\sum_{p_k \in \mathcal{P}} \exp(\mathbf{u}_i^T \cdot \mathbf{P} \cdot CNN(p_k))} \qquad (1.10)$$

where f_{is} denotes if p_s is posted by u_i, $CNN(p_s)$ is the extracted feature from the image p_s via CNN (implemented by VGG-16 [48]), $\mathbf{P} \in R^{K \times d}$ is the interaction matrix between the visual contents and latent user features, and d is the dimension of the visual contents.

Similarly, the probability $P(g_{jt} = 1|l_j, p_t)$ is defined to measure whether the image p_t is tagged to the POI l_j for learning the maximum likelihood on $\mathcal{P}_l$. Considering an image p_t associated with POI l_j, $CNN(P_t)$ denotes the visual feature of image p_t, and $\mathbf{l}_j \in R^K$ denotes the latent feature of POI l_j. Hence, the probability of $P(g_{jt} = 1|l_j, p_t)$ is defined as follows,

$$P(g_{jt} = 1|l_j, p_t) = \frac{\exp(\mathbf{l}_j^T \cdot \mathbf{Q} \cdot CNN(p_t))}{\sum_{p_k \in \mathcal{P}} \exp(\mathbf{l}_j^T \cdot \mathbf{Q} \cdot CNN(p_k))}, \qquad (1.11)$$

where g_{jt} denotes if photo p_t is tagged to POI l_j, $\mathbf{Q} \in R^{K \times d}$ is the interaction matrix between the visual contents and latent POI features, and d is the dimension of the visual contents.

Given the check-in matrix $\mathbf{G}$, user images $\mathcal{P}_u$, and POI images $\mathcal{P}_l$, the visual feature enhanced POI recommendation framework is defined via maximizing the following logarithmic posterior distribution,

$$\max_{\mathbf{U},\mathbf{L},\mathbf{P},\mathbf{Q}} \log P(\mathbf{U}, \mathbf{L}, \mathbf{P}, \mathbf{Q} | \mathbf{G}, \mathscr{P}_u, \mathscr{P}_l), \tag{1.12}$$

where $\mathbf{U}, \mathbf{L}, \mathbf{P}, \mathbf{Q}$ are user latent feature matrix, POI latent feature matrix, user-photo feature interaction matrix, and POI-photo feature interaction matrix, respectively. Furthermore, based on $P(f_{is} = 1|u_i, p_s)$ and $P(g_{jt} = 1|l_j, p_t)$ this posterior distribution can be learned through the maximum likelihood over $\mathbf{G}$, $\mathscr{P}_u$, and $\mathscr{P}_l$ with regularizations,

$$\max_{\mathbf{U},\mathbf{L},\mathbf{P},\mathbf{Q}} \alpha \left(\sum_{i=1}^{|\mathscr{U}|} \sum_{p_k \in \mathscr{P}_{u_i}} \log P(f_{ik} = 1|u_i, p_k) + \sum_{j=1}^{|\mathscr{L}|} \sum_{p_k \in \mathscr{P}_{l_j}} \log P(g_{jk} = 1|l_j, p_k)) \right.$$
$$-||\mathbf{Y} \odot (\mathbf{G} - \mathbf{U}^T\mathbf{L}||_F^2 - \lambda_1(||\mathbf{U}||_F^2 + ||\mathbf{L}||_F^2) - \lambda_2(||\mathbf{P}||_F^2 + ||\mathbf{Q}||_F^2), \tag{1.13}$$

where $\mathbf{Y}$ is the indicator matrix that constrains the calculation only valid for the non-zero entries in $\mathbf{G}$, $\odot$ is the Hadamard product, α is the hyperparamter to constrain the effect of visual modeling, and λ_1 and λ_2 are regularizations to avoid overfitting. After learning the objective function Eq. (1.13), the top k POIs can be selected according to the value of $\mathbf{U}^T\mathbf{L}$ that measures the user check-in preference on POIs.

1.2.3 Methodologies

Based on the analysis of user check-in behaviors, researchers propose a variety of POI recommendation systems, which can be categorized into two types: fused model and joint model. The fused model fuses recommended results from collaborative filtering method and recommended results from models capturing geographical influence, social influence, and temporal influence. The joint model establishes a joint model to learn the user preference and the influential factors together.

1.2.3.1 Fused Model

The fused model establishes a model for each influential factor and combines their recommended results with suggestions from the collaborative filtering model [40] that captures user preference on POIs. Since social influence provides limited improvements in POI recommendation and user comments are usually missing in users' check-ins, geographical influence and temporal influence constitute two important factors for POI recommendation. Hence, a typically fused model [3, 57, 68] recommends POIs through combining the traditional collaborative filtering methods and influential factors, especially including geographical influence or temporal influence. Using collaborative filtering methods to capture the user preference can be categorized into two types: memory-based method (e.g., user-based) and model-based method (matrix factorization).

Representative Work for Memory-based Method

In [57], Ye et al. propose a fused framework for POI recommendation, which captures the user preference, social influence, and geographical influence. Specifically, Ye et al. [57] use the user-based collaborative filtering to model the user preference, friend-based collaborative filtering for social influence, power law distribution model for geographical model. Let $S_{i,j}$ denote the check-in probability score of user u_i at POI l_j. $S_{i,j}^u$, $S_{i,j}^s$, and $S_{i,j}^g$ denote the check-in probability scores of user u_i at POI l_j, corresponding to recommendation results based on user preference, social influence, and geographical influence, respectively. Then, the fused recommendation result is formulated as,

$$S_{i,j} = (1 - \alpha - \beta)S_{i,j}^u + \alpha S_{i,j}^s + \beta S_{i,j}^g, \tag{1.14}$$

where the two weighting parameters α and β ($0 \le \alpha + \beta \le 1$) denote the relative importance of social influence and geographical influence comparing to user preference.

Specifically, $S_{i,j}^u$, $S_{i,j}^s$, and $S_{i,j}^g$ are obtained from the check-in probability $p_{i,j}^u$, $p_{i,j}^s$, and $p_{i,j}^g$ for a user u_i to visit a POI l_j. $p_{i,j}^s$, and $p_{i,j}^g$ can be calculated using Eq. (1.7) described in Sect. 1.2.2.3 and Eq. (1.2) described in Sect. 1.2.2.1, respectively. $p_{i,j}^u$ is calculated through the user-based CF model,

$$p_{i,j}^u = \frac{\sum_{u_k} w_{i,k} \cdot c_{k,j}}{\sum_{u_k} w_{i,k}}, \tag{1.15}$$

where $c_{k,j}$ denotes the check-in frequency of user u_k at POI l_j, $w_{i,k}$ is the similarity weight between user u_i and u_k calculated via the Pearson Correlation Covariance [40]. After obtaining the check-in probability estimation, the corresponding scores are as follows,

$$S_{i,j}^u = \frac{p_{i,j}^u}{Z_i^u}, \quad S_{i,j}^s = \frac{p_{i,j}^s}{Z_i^s}, \quad S_{i,j}^g = \frac{p_{i,j}^g}{Z_i^g}, \tag{1.16}$$

where Z_i^u, Z_i^s, Z_i^g are normalization terms. $Z_i^u = \max_{l_j \in L \setminus L_i}\{p_{i,j}^u\}$, $Z_i^s = \max_{l_j \in L \setminus L_i}\{p_{i,j}^s\}$, $Z_i^g = \max_{l_j \in L \setminus L_i}\{p_{i,j}^g\}$, where $L \setminus L_i$ denotes the POIs user u_i has not visited.

Representative Work for Model-based Method

In [3], Cheng et al. employ probabilistic matrix factorization (PMF) [42] and probabilistic factor model (PFM) [31] to learn user preference for recommending POIs. Suppose $\mathscr{U}$ denote the set of users and $\mathscr{L}$ denote the set of POIs. U_i and L_j denote the latent feature of user u_i and POI l_j. PMF-based method assumes Gaussian distribution on observed check-in data and Gaussian priors on the user latent feature matrix U and POI latent feature matrix L. Then, the objective function to learn the model is as follows,

$$\min_{U,L} \sum_{i=1}^{|\mathcal{U}|} \sum_{j=1}^{|\mathcal{L}|} I_{ij}(g(c_{ij}) - g(U_i^T L_j))^2 + \lambda_1 ||U||_F^2 + \lambda_2 ||L||_F^2, \quad (1.17)$$

where $g(x) = \frac{1}{1+e^{-x}}$ is the sigmoid function, c_{ij} is the checked-in frequency of user u_i at POI l_j. I_{ij} is the indicator function to record the check-in state of u_i at l_j. Namely, I_{ij} equals one when the i-th user has checked-in at j-th POI; otherwise zero. After learning the user and POI latent features, the preference score of u_i over l_j is measured by the following score function,

$$P(F_{ul}) = g(U_i^T L_j), \quad (1.18)$$

where $g(\cdot)$ is the sigmoid function.

In addition, the geographical influence can be modeled through MGM, shown in Eq. (1.3) of Sect. 1.2.2.1. Then, a fused model is proposed to combine user preference learned from Eq. (1.17) and geographical influence modeled in Eq. (1.3). The proposed model determines the probability P_{ul} of a user u visiting a location l via the product of the preference score estimation and the probability of whether a user will visit that place in terms of geographical influence,

$$P_{ul} = P(F_{ul}) \cdot P(l|C_u), \quad (1.19)$$

where $P(l|C_u)$ is calculated via the MGM and $P(F_{ul})$ encodes a user's preference on a location.

1.2.3.2 Joint Model

Different from the fused model, the joint model learns several influential factors together and then recommends POIs from the jointly learned model. Compared with the fused model, a joint model connects different influential factors into the same final training target—the check-in behavior. The joint model depicts the check-in behavior as a synthesized decision influenced by several factors together, which better reflects the real scenario than the fused model. This advantage over the fused model makes the joint model attract more attention. Recently a number of joint models [9, 10, 17, 21, 24, 25, 30, 54, 61] have been proposed for POI recommendation, which can be categorized into three types: (1) MF-based joint model that incorporates factors such as geographical influence and temporal influence into traditional collaborative filtering model like matrix factorization and tensor factorization, e.g., [9, 10, 24, 30, 54]; (2) generative graphical model that establishes a generative model according to the check-ins and extra influences like geographical information, e.g., [17, 21, 25, 61]; (3) neural network model that jointly models the influential factors in a neural network, e.g., [28, 29].

Representative Work for MF-based Joint Model

This section reports two representative studies about the MF-based joint model, which incorporate temporal effect and geographical effect into a matrix factorization framework, respectively.

In [9], Gao et al. propose a Location Recommendation framework with Temporal effects (LRT), which incorporates temporal influence into a matrix factorization model. The LRT model contains two assumptions on temporal effect: (1) non-uniformness, users' check-in preferences change at different hours of one day; (2) consecutiveness, users' check-in preferences are similar in consecutive time slots. To model the non-uniformness, LRT separates a day into T slots, and defines time-dependent user latent feature $U_t \in R^{m \times d}$, where m is the number of users, d is the latent feature dimension, and $t \in [1, T]$ indexes time slots. Suppose that $C_t \in R^{m \times n}$ denotes a matrix depicting the check-in frequency at temporal state t. U and L denote the latent feature matrix for user and POI, respectively. Using the non-negative matrix factorization to model the POI recommendation system, the time-dependent objective function is as follows,

$$\min_{U_t \geq 0, L \geq 0} \sum_{t=1}^{T} ||Y_t \odot (C_t - U_t L^T)||_F^2 + \alpha \sum_{t=1}^{T} ||U_t||_F^2 + \beta ||L||_F^2, \quad (1.20)$$

where Y_t is the corresponding indicator matrix, α and β are the regularizations. Furthermore, the temporal consecutiveness inspires to minimize the following term,

$$\min \sum_{t=1}^{T} \sum_{i=1}^{m} \phi_i(t, t-1) ||U_t(i, :) - U_{t-1}(i, :)||_2^2, \quad (1.21)$$

where $\phi_i(t, t-1) \in [0, 1]$ is defined as a temporal coefficient that measures user preference similarity between temporal state t and $t-1$. The temporal coefficient could be calculated via cosine similarity according to users' check-ins at state t and $t-1$. To represent the Eq. (1.21) in matrix form, we get

$$\min \sum_{t=1}^{T} Tr((U_t - U_{t-1})^T \Sigma_t (U_t - U_{t-1})), \quad (1.22)$$

where $\Sigma_t \in R^{m \times m}$ is the diagonal temporal coefficient matrix among m users. Combining the two minimization targets, the objective function of the LRT model is gained as follows,

$$\min_{U_t \geq 0, L \geq 0} \sum_{t=1}^{T} ||Y_t \odot (C_t - U_t L^T)||_F^2 + \alpha \sum_{t=1}^{T} ||U_t||_F^2 + \beta ||L||_F^2$$

$$+\lambda \sum_{t=1}^{T} Tr((U_t - U_{t-1})^T \Sigma_t (U_t - U_{t-1})), \tag{1.23}$$

where λ is a non-negative parameter to control the temporal regularization. User and location latent representations can be learned by solving the above optimization problem. Then, the user check-in preference $\hat{C}_t(i,j)$ at each temporal state can be estimated by the product of user latent feature and location feature ($U_t(i,:)L(j,:)^T$). Recommending POIs for users is to find POIs with the higher value of $\hat{C}(i,j)$. To aggregate different temporal states' contributions, $\hat{C}(i,j)$ is estimated through

$$\hat{C}(i,j) = f(\hat{C}_1(i,j), \hat{C}_2(i,j), \dots, \hat{C}_T(i,j)), \tag{1.24}$$

where $f(\cdot)$ is an aggregation function, e.g., sum, mean, maximum, and voting operation.

In [24], Lian et al. propose the Geo matrix factorization (GeoMF) model to incorporate geographical influence into a weighted regularized matrix factorization model (WRMF) [18, 36]. WRMF is a popular model for one-class collaborative filtering problem, learning implicit feedback for recommendations. GeoMF treats the user check-in as implicit feedback and leverages a 0/1 rating matrix to represent the user check-ins. Furthermore, GeoMF employs an augmented matrix to recover the rating matrix, as shown in Fig. 1.12. Each entry in the rating matrix is the combination of two interactions: user feature and POI feature, users' activity area representation and POIs' influence area representation. Suppose there are m users and n POIs. The latent feature dimension is d for user and POI representations, and the latent feature dimension is l for users' activity area and POIs' influence area representations. Then the estimated rating matrix can be formulated as,

$$\tilde{R} = PQ^T + XY^T, \tag{1.25}$$

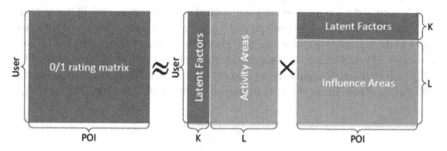

Fig. 1.12 Demonstration of GeoMF model [24]

where $\tilde{R} \in R^{m \times n}$ is the estimated matrix, $P \in R^{m \times d}$ and $Q \in R^{n \times d}$ are the user latent matrix and POI latent matrix, respectively. In addition, $X \in R^{m \times l}$ and $Y \in R^{n \times l}$ are user activity area representation matrix and POI activity area representation matrix, respectively. Define W as the binary weighted matrix whose entry w_{ui} is set as follows,

$$w_{ui} = \begin{cases} \alpha(c_{ui}) + 1 & \text{if } c_{ui} > 0 \\ 1 & \text{otherwise,} \end{cases} \quad (1.26)$$

where c_{ui} is user u's check-in frequency at POI l_i, $\alpha(c_{ui}) > 0$ is a monotonically increasing function with respect to c_{ui}. Following the scheme of WRMF model, the objective function of GeoMF is formulated as,

$$\underset{P,Q,X}{\arg \min} ||W \odot (R - PQ^T - XY^T)||_F^2 + \gamma (||P||_F^2 + ||Q||_F^2) + \lambda ||X||_1, \quad (1.27)$$

where Y is POIs' influence area matrix generated from a Gaussian kernel function, P, Q, and X are parameters that need to learn, and γ and λ are regularizations. After learning the latent features from Eq. (1.27), the proposed model estimates the check-in possibility according to Eq. (1.25), and then recommends the POIs with higher values for each user.

Representative Work for Generative Graphical Model

This section presents the representative research about the generative graphical model, which incorporates geographical influence into a generative graphical model.

In [25], Liu et al. propose a geographical probabilistic factor analysis framework that takes various factors into consideration, including user preferences, the geographical influence, and the user mobility pattern. The proposed model mimics the user check-in decision process to learn geographical user preferences for effective POI recommendations. Figure 1.13 demonstrates the graphical representation of the proposed model. Specifically, the proposed model assumes that the geographical locations have been clustered into several latent regions denoted as R. A multinomial distribution is applied to model user mobility over the regions R, $r \sim p(r|\eta_u)$, where η_u is a user dependent distribution over latent regions for user u_i. Then, each region $r \in R$ is assumed to be a Gaussian geographical distribution and the POI l_j is characterized by $l \sim \mathcal{N}(\mu_r, \sum_r)$ with μ_r and $\sum_r$ being the mean vector and covariance matrix of the region. In addition, the user check-in process is affected by the following factors: (1) each user u_i is associated with an interest $\alpha(i, j)$ with respect to POI l_j; (2) each POI l_j has popularity ρ_j; and (3) the distance between the user and the POI $d(u_i, l_j)$. Then, the probability of user u_i visiting POI l_j can be formulated as,

$$p(u_i, l_j) \propto \alpha(i, j)\rho_j(d_0 + d(u_i, l_j))^{-\tau}, \quad (1.28)$$

where a power-law like the parametric term $(d_0 + d(u_i, l_j))^{-\tau}$ is used to model the distance factor. Moreover, the user preference for POI can be represented as a lin-

Fig. 1.13 A graphical
representation of the
model [25]

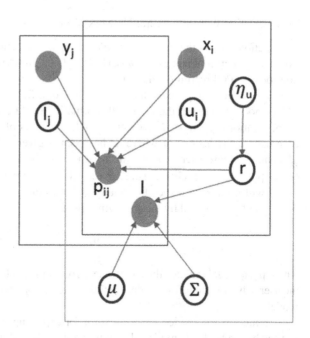

Algorithm 1: Model generative process

1: Draw a region $r \sim$ Multinomial(η_u)
2: Draw a location $l \sim \mathcal{N}(\mu_r, \Sigma_r)$
3: Draw a user preference
4:     Generate user latent factor $\mathbf{u}_i \sim P(u_i; \Phi_{\mathbf{u}})$
5:     Generate POI latent factor $\mathbf{l}_j \sim P(\mathbf{l}_j; \Phi_{\mathbf{l}_j})$
6:     User-item preference $\alpha(i,j) = \mathbf{u}_i^T \mathbf{l}_i + x_i^T W y_j$
7: Generate $p_{ij} \sim P(f_{ij})$, where $p_{ij} = (\mathbf{u}_i^T \mathbf{l}_j + x_i^T W y_j)\rho_j(d_0 + d(u_i, l_j))^{-\tau}$

---

ear combination of a latent factor $\mathbf{u}_i^T \mathbf{l}_j$ and a function of user and item observable
properties $x_i^T W y_j$, namely

$$\alpha(i,j) = \mathbf{u}_i^T \mathbf{l}_j + x_i^T W y_j. \tag{1.29}$$

The proposed model uses implicit user check-in data to model user preferences and
the distribution of check-in counts are usually skewed, so a Bayesian probabilistic
non-negative latent factor model is employed: $p_{ij} \sim P(f_{ij})$ where $f_{ij} = \alpha(i,j)\rho_j(d_0 + d(u_i, l_j))^{-\tau}$. The proposed model shown in Fig. 1.13 can be generated according to
Algorithm 1.

After the parameters are learned, the proposed model predicts the number
of check-ins of a user for a given POI as $\mathbb{E}(p_{ij}|u_i, l_j) = (\mathbf{u}_i^T \mathbf{l}_j + x_i^T W y_j)\rho_j(d_0 + d(u_i, l_j))^{-\tau}$. Moreover, POI recommendations are based on the predicted check-in times. The larger the predicted value is, the more likely the user will choose this
POI.

Representative Work for Neural Network Model

This section reports the representative research about neural network model, which extends the recurrent neural network (RNN) [15] with spatial and temporal information for next POI recommendation.

In [28], Liu et al. propose the Spatial Temporal Recurrent Neural Network (ST-RNN) model to predict the next POI. Formally, let P be a set of users and Q be a set of locations, $\mathbf{p}_u, \mathbf{q}_l \in R^d$ indicate the latent vectors of user u and location l. For each user u, the history of where he has been is given as $Q^u = q^u_{t_1}, q^u_{t_2}, \ldots$, where $q^u_{t_i}$ denotes where user u is at time t_i. And the history of all users is denoted as $Q^U = \{Q^{u1}, Q^{u2}, \ldots\}$. Given historical records of a users, the task is to predict where a user will go next at a specific time t. The possibility of user u visit location l at time t can be estimated by the following function,

$$o_{u,t,l} = (\mathbf{h}^u_{t,q_l} + \mathbf{p}_u)^T \mathbf{q}_l, \qquad (1.30)$$

where $\mathbf{p}_u, \mathbf{q}_l \in R^d$ indicate the latent vectors of user u and location l, and $\mathbf{h}^u_{t,q_l}$ captures the user's dynamic interests under spatial and temporal contexts learned from a RNN model.

Specifically, the hidden representation for capturing user interest's dynamics is learned by the ST-RNN model, shown in Fig. 1.14. $\mathbf{h}^u_{t,q^u_t}$ denotes the dynamic interest representation of user u at time t, $\mathbf{q}^u_{t_i}$ is the latent vector of the location the user visits at time t_i, and $\hat{w}$ is window size. Therefore, $\mathbf{h}^u_{t,q^u_t}$ can be formulated from the visited POIs in the watching window,

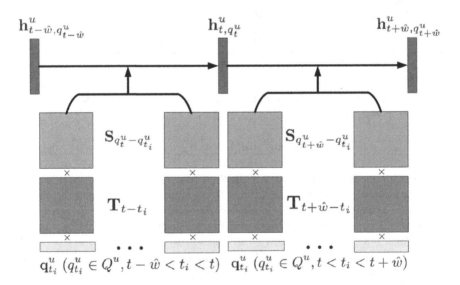

Fig. 1.14 Overview of ST-RNN [28]

$$\mathbf{h}_{t,q_t^u}^u = f\Big(\sum_{q_{t_i}^u \in Q^u, t-\hat{w}<t_i<t} \mathbf{M}\mathbf{q}_{t_i}^u + \mathbf{C}\mathbf{h}_{t-\hat{w},q_{t-\hat{w}}^u}\Big), \qquad (1.31)$$

where $f(x)$ is sigmoid function, $\mathbf{M}$ denotes the transition matrix for input elements to capture the current behavior of the user, and $\mathbf{C}$ is the recurrent connection of the previous status propagating sequential signals. Furthermore, $\mathbf{M}$ is used to capture the spatial and temporal contexts, defined as $\mathbf{M} = \mathbf{S}_{q_t^u - q_{t_i}^u} \mathbf{T}_{t-t_i}$, where $\mathbf{S}_{q_t^u - q_{t_i}^u}$ is the distance-specific transition matrix for the geographical distance between q_t^u and $q_{t_i}^u$, and $\mathbf{T}_{t-t_i}$ denotes the time-specific transition matrix for the time interval $t - t_i$. To learn the parameters $\mathbf{S}_{q_t^u - q_{t_i}^u}$, $\mathbf{T}_{t-t_i}$, $\mathbf{q}_{t_i}^u$, $\mathbf{C}$, $\mathbf{h}_{t,q_t^u}^u$, $\mathbf{p}_u$, and $\mathbf{q}_l$, [28] uses the Bayesian Personalized Ranking (BPR) [37] and Back Propagation Through Time (BPTT) [41] to infer the model.

1.3 Book Organization

The remainder of this book is organized as follows.

- Chapter 2
  This chapter attempts to understand the user mobility in LBSNs from the geographical perspective and capture the geographical influence for POI recommendation. Section 2.2 introduces the related work. Section 2.3 demonstrates the two models: Gaussian Mixture Model (GMM) and genetic algorithm based GMM (GA-GMM) to capture the geographical influence for POI recommendation. Section 2.4 compares our proposed methods with state-of-the-art geographical models. Experimental results show that our proposed models better capture the geographical influence and perform better for POI recommendation. Section 2.5 summarizes this chapter and draws the conclusion.
- Chapter 3
  This chapter analyzes the user mobility from the temporal perspective and aim to capture the temporal influence for the POI recommendation. Section 3.2 reviews the most relevant work. Section 3.3 introduces our empirical analysis on check-in data and demonstrates the time labeling scheme. Next, Section 3.4 presents the details of the proposed Aggregated Temporal Tensor Factorization (ATTF) model and understands the proposed model from the neural network perspective. Then, Sect. 3.5 reports experimental results conducted on two real-world datasets. Finally, Sect. 3.6 summarizes this chapter and draws the conclusion.
- Chapter 4
  This chapter proposes the **Geo-Te**mporal **s**equential **e**mbedding **r**ank (Geo-Teaser) model for POI recommendation to capture the user preference, check-ins' sequential pattern, and the user spatial and temporal mobility pattern. Section 4.2 reviews the related work. Section 4.3 introduces two real-world datasets and reports empirical data analysis that motivates our method. Next, Sect. 4.4 introduces our

proposed Geo-Teaser model and show the learning algorithm. Then, Sect. 4.5 evaluates our proposed model. Finally, Sect. 4.6 concludes this chapter.

- Chapter 5
  This chapter proposes the spatial-temporal latent ranking (STELLAR) system to resolve the time-aware successive POI recommendation problem. Section 5.2 reviews the most relevant work and summarizes the connections of the proposed model and prior work. Section 5.3 introduces the empirical analysis on check-in data to show the spatial and temporal properties. Next, we Section 5.4 presents the details of our model and show the learning procedures. Then, Sect. 5.5 reports experimental results conducted on two real-world datasets. Finally, Sect. 5.6 concludes this chapter.
- Chapter 6
  This chapter summarizes this book and points out the future work direction. In particular, Sect. 6.1 draws the conclusion of this book. Then, Sect. 6.2 points out the future work direction in three aspects: ranking based model, online recommendation, and deep learning based recommendation.

References

1. Bhargava, P., Phan, T., Zhou, J., Lee, J.: Who, What, When, and Where: multi-dimensional collaborative recommendations using tensor factorization on sparse user-generated data. In: Proceedings of the 24th International Conference on World Wide Web, pp. 130–140. ACM (2015)
2. Brockmann, D., Hufnagel, L., Geisel, T.: The scaling laws of human travel. Nature **439**(7075), 462–465 (2006)
3. Cheng, C., Yang, H., King, I., Lyu, M.R.: Fused matrix factorization with geographical and social influence in location-based social networks. In: Proceedings of the Twenty-Sixth AAAI Conference on Artificial Intelligence, pp. 17–23. AAAI Press (2012)
4. Cheng, C., Yang, H., Lyu, M.R., King, I.: Where you like to go next: successive point-of-interest recommendation. In: Proceedings of the Twenty-Third International Joint Conference on Artificial Intelligence, pp. 2605–2611. AAAI Press (2013)
5. Cheng, Z., Caverlee, J., Lee, K., Sui, D.Z.: Exploring millions of footprints in location sharing services. In: Fifth International AAAI Conference on Weblogs and Social Media, pp. 81–88. AAAI Press (2011)
6. Cho, E., Myers, S.A., Leskovec, J.: Friendship and mobility: user movement in location-based social networks. In: Proceedings of the 17th ACM SIGKDD international conference on Knowledge discovery and data mining, pp. 1082–1090. ACM (2011)
7. Feng, S., Li, X., Zeng, Y., Cong, G., Chee, Y.M., Yuan, Q.: Personalized ranking metric embedding for next new POI recommendation. In: Proceedings of the 24th International Conference on Artificial Intelligence, pp. 2069–2075. AAAI Press (2015)
8. Ference, G., Ye, M., Lee, W.C.: Location recommendation for out-of-town users in location-based social networks. In: Proceedings of the 22nd ACM International Conference on Information and Knowledge Management, pp. 721–726. ACM (2013)
9. Gao, H., Tang, J., Hu, X., Liu, H.: Exploring temporal effects for location recommendation on location-based social networks. In: Proceedings of the 7th ACM Conference on Recommender Systems, pp. 93–100. ACM (2013)

10. Gao, H., Tang, J., Hu, X., Liu, H.: Content-aware point-of-interest recommendation on location-based social networks. In: Proceedings of the Twenty-Ninth AAAI Conference on Artificial Intelligence, pp. 1721–1727. AAAI Press (2015)

11. Gao, H., Tang, J., Liu, H.: Exploring social-historical ties on location-based social networks. In: Sixth International AAAI Conference on Weblogs and Social Media. AAAI Press (2012)

12. Gao, H., Tang, J., Liu, H.: gSCorr: modeling geo-social correlations for new check-ins on location-based social networks. In: Proceedings of the 21st ACM International Conference on Information and Knowledge Management, pp. 1582–1586. ACM (2012)

13. Gao, H., Tang, J., Liu, H.: Addressing the cold-start problem in location recommendation using geo-social correlations. Data Min. Knowl. Discov. 29(2), 299–323 (2015)

14. Gonzalez, M.C., Hidalgo, C.A., Barabasi, A.L.: Understanding individual human mobility patterns. Nature 453(7196), 779–782 (2008)

15. Goodfellow, I., Bengio, Y., Courville, A.: Deep Learning. MIT Press (2016)

16. He, J., Li, X., Liao, L., Song, D., Cheung, W.K.: Inferring a personalized next point-of-interest recommendation model with latent behavior patterns. In: Thirtieth AAAI Conference on Artificial Intelligence, pp. 137–143 (2016)

17. Hu, B., Ester, M.: Social topic modeling for point-of-interest recommendation in location-based social networks. In: 2014 IEEE International Conference on Data Mining, pp. 845–850. IEEE (2014)

18. Hu, Y., Koren, Y., Volinsky, C.: Collaborative filtering for implicit feedback datasets. In: 2008 Eighth IEEE International Conference on Data Mining, pp. 263–272. IEEE (2008)

19. Jamali, M., Ester, M.: Trustwalker: a random walk model for combining trust-based and item-based recommendation. In: Proceedings of the 15th ACM SIGKDD International Conference on Knowledge Discovery and Data Mining, pp. 397–406. ACM (2009)

20. Jamali, M., Ester, M.: A matrix factorization technique with trust propagation for recommendation in social networks. In: Proceedings of the Fourth ACM Conference on Recommender Systems, pp. 135–142. ACM (2010)

21. Kurashima, T., Iwata, T., Hoshide, T., Takaya, N., Fujimura, K.: Geo topic model: joint modeling of user's activity area and interests for location recommendation. In: Proceedings of the Sixth ACM International Conference on Web Search and Data Mining, pp. 375–384. ACM (2013)

22. Li, H., Ge, Y., Hong, R., Zhu, H.: Point-of-interest recommendations: learning potential check-ins from friends. In: Proceedings of the 22nd ACM SIGKDD International Conference on Knowledge Discovery and Data Mining, pp. 975–984. ACM (2016)

23. Lian, D., Ge, Y., Zhang, F., Yuan, N.J., Xie, X., Zhou, T., Rui, Y.: Content-aware collaborative filtering for location recommendation based on human mobility data. In: 2015 IEEE International Conference on Data Mining (ICDM), pp. 261–270. IEEE (2015)

24. Lian, D., Zhao, C., Xie, X., Sun, G., Chen, E., Rui, Y.: GeoMF: Joint geographical modeling and matrix factorization for point-of-interest recommendation. In: ACM SIGKDD International Conference on Knowledge Discovery and Data Mining, pp. 831–840. ACM (2014)

25. Liu, B., Fu, Y., Yao, Z., Xiong, H.: Learning geographical preferences for point-of-interest recommendation. In: Proceedings of the 19th ACM SIGKDD International Conference on Knowledge Discovery and Data Mining, pp. 1043–1051. ACM (2013)

26. Liu, B., Xiong, H.: Point-of-interest recommendation in location based social networks with topic and location awareness. In: Siam International Conference on Data Mining, pp. 396–404. SIAM (2013)

27. Liu, B., Xiong, H., Papadimitriou, S., Fu, Y., Yao, Z.: A general geographical probabilistic factor model for point of interest recommendation. IEEE Trans. Knowl. Data Eng. 27(5), 1167–1179 (2015)

28. Liu, Q., Wu, S., Wang, L., Tan, T.: Predicting the next location: a recurrent model with spatial and temporal contexts. In: Thirtieth AAAI Conference on Artificial Intelligence, pp. 194–200 (2016)

29. Liu, X., Liu, Y., Li, X.: Exploring the context of locations for personalized location recommendations. In: Proceedings of the Twenty-Fifth International Joint Conference on Artificial Intelligence, pp. 1188–1194. AAAI Press (2016)

30. Liu, Y., Wei, W., Sun, A., Miao, C.: Exploiting geographical neighborhood characteristics for location recommendation. In: ACM International Conference on Conference on Information and Knowledge Management, pp. 739–748. ACM (2014)

31. Ma, H., Liu, C., King, I., Lyu, M.R.: Probabilistic factor models for web site recommendation. In: Proceedings of the 34th International ACM SIGIR Conference on Research and Development in Information Retrieval, pp. 265–274. ACM (2011)

32. Ma, H., Yang, H., Lyu, M.R., King, I.: Sorec: social recommendation using probabilistic matrix factorization. In: Proceedings of the 17th ACM Conference on Information and Knowledge Management, pp. 931–940. ACM (2008)

33. Ma, H., Zhou, D., Liu, C., Lyu, M.R., King, I.: Recommender systems with social regularization. In: Proceedings of the Fourth ACM International Conference on Web Search and Data Mining, pp. 287–296. ACM (2011)

34. Massa, P., Avesani, P.: Trust-aware recommender systems. In: Proceedings of the 2007 ACM Conference on Recommender Systems, pp. 17–24. ACM (2007)

35. Noulas, A., Scellato, S., Lathia, N., Mascolo, C.: Mining user mobility features for next place prediction in location-based services. In: 2012 IEEE 12th International Conference on Data Mining, pp. 1038–1043. IEEE (2012)

36. Pan, R., Zhou, Y., Cao, B., Liu, N.N., Lukose, R., Scholz, M., Yang, Q.: One-class collaborative filtering. In: Proceedings of the 2008 Eighth IEEE International Conference on Data Mining, pp. 502–511. IEEE Computer Society (2008)

37. Rendle, S., Balby Marinho, L., Nanopoulos, A., Schmidt-Thieme, L.: Learning optimal ranking with tensor factorization for tag recommendation. In: Proceedings of the 15th ACM SIGKDD International Conference on Knowledge Discovery and Data Mining, pp. 727–736. ACM (2009)

38. Rendle, S., Freudenthaler, C., Schmidt-Thieme, L.: Factorizing personalized markov chains for next-basket recommendation. In: Proceedings of the 19th International Conference on World Wide Web, pp. 811–820. ACM (2010)

39. Rhee, I., Shin, M., Hong, S., Lee, K., Kim, S.J., Chong, S.: On the levy-walk nature of human mobility. IEEE/ACM Trans. Netw. (TON) **19**(3), 630–643 (2011)

40. Ricci, F., Rokach, L., Shapira, B.: Introduction to recommender systems handbook. Springer (2011)

41. Rumelhart, D.E., Hinton, G.E., Williams, R.J.: Learning representations by back-propagating errors. MIT Press (1986)

42. Salakhutdinov, R., Mnih, A.: Probabilistic matrix factorization. Adv. Neural Inf. Process. Syst. **20**, 1257–1264 (2008)

43. Sang, J., Mei, T., Xu, C.: Activity sensor: check-in usage mining for local recommendation. ACM Trans. Intell. Syst. Technol. (TIST) **6**(3), 41 (2015)

44. Sattari, M., Manguoglu, M., Toroslu, I.H., Symeonidis, P., Senkul, P., Manolopoulos, Y.: Geo-activity recommendations by using improved feature combination. In: Proceedings of the 2012 ACM Conference on Ubiquitous Computing, pp. 996–1003. ACM (2012)

45. Scellato, S., Mascolo, C., Musolesi, M., Latora, V.: Distance matters: geo-social metrics for online social networks. In: Proceedings of the 3rd Wonference on Online Social Networks, pp. 8–8. USENIX Association (2010)

46. Scellato, S., Noulas, A., Mascolo, C.: Exploiting place features in link prediction on location-based social networks. In: ACM SIGKDD International Conference on Knowledge Discovery and Data Mining, San Diego, CA, USA, August, pp. 1046–1054. ACM (2011)

47. Silverman, B.W.: Density estimation for statistics and data analysis, vol. 26. CRC press (1986)

48. Simonyan, K., Zisserman, A.: Very deep convolutional networks for large-scale image recognition (2014). arXiv:1409.1556

49. Wang, S., Wang, Y., Tang, J., Shu, K., Ranganath, S., Liu, H.: What your images reveal: exploiting visual contents for point-of-interest recommendation. In: Proceedings of the 26th International Conference on World Wide Web, pp. 391–400. International World Wide Web Conferences Steering Committee (2017)

50. Wang, W., Yin, H., Chen, L., Sun, Y., Sadiq, S., Zhou, X.: St-sage: a spatial-temporal sparse additive generative model for spatial item recommendation. ACM Trans. Intell. Syst. Technol. (TIST) **8**(3), 48 (2017)

51. Wang, X., Zhao, Y.L., Nie, L., Gao, Y., Nie, W., Zha, Z.J., Chua, T.S.: Semantic-based location recommendation with multimodal venue semantics. IEEE Trans. Multimed. **17**(3), 409–419 (2015)

52. Wang, Y., Yuan, N.J., Lian, D., Xu, L., Xie, X., Chen, E., Rui, Y.: Regularity and conformity: location prediction using heterogeneous mobility data. In: ACM SIGKDD International Conference on Knowledge Discovery and Data Mining, pp. 1275–1284. ACM (2015)

53. Wang, Z.S., Juang, J.F., Teng, W.G.: Predicting POI visits with a heterogeneous information network. In: 2015 Conference on Technologies and Applications of Artificial Intelligence (TAAI), pp. 388–395. IEEE (2015)

54. Yang, D., Zhang, D., Yu, Z., Wang, Z.: A sentiment-enhanced personalized location recommendation system. In: Proceedings of the 24th ACM Conference on Hypertext and Social Media, pp. 119–128. ACM (2013)

55. Yang, D., Zhang, D., Zheng, V.W., Yu, Z.: Modeling user activity preference by leveraging user spatial temporal characteristics in LBSNs. IEEE Trans. Syst. Man Cybern. Syst. **45**(1), 129–142 (2015)

56. Ye, M., Yin, P., Lee, W.C.: Location recommendation for location-based social networks. In: Proceedings of the 18th SIGSPATIAL International Conference on Advances in Geographic Information Systems, pp. 458–461. ACM (2010)

57. Ye, M., Yin, P., Lee, W.C., Lee, D.L.: Exploiting geographical influence for collaborative point-of-interest recommendation. In: Proceedings of the 34th International ACM SIGIR Conference on Research and Development in Information Retrieval, pp. 325–334. ACM (2011)

58. Yin, H., Cui, B., Chen, L., Hu, Z., Zhou, X.: Dynamic user modeling in social media systems. ACM Trans. Inf. Syst. (TOIS) **33**(3), 10 (2015)

59. Yin, H., Cui, B., Sun, Y., Hu, Z., Chen, L.: LCARS: a spatial item recommender system. ACM Trans. Inf. Syst. (TOIS) **32**(3), 11 (2014)

60. Yin, H., Hu, Z., Zhou, X., Wang, H., Zheng, K., Nguyen, Q.V.H., Sadiq, S.: Discovering interpretable geo-social communities for user behavior prediction. In: The 32nd IEEE International Conference on Data Engineering, pp. 942–953. IEEE (2016)

61. Yin, H., Sun, Y., Cui, B., Hu, Z., Chen, L.: Lcars: A location-content-aware recommender system. In: ACM SIGKDD International Conference on Knowledge Discovery and Data Mining, pp. 221–229. ACM (2013)

62. Yin, H., Wang, W., Wang, H., Chen, L., Zhou, X.: Spatial-aware hierarchical collaborative deep learning for POI recommendation. IEEE Trans. Knowl. Data Eng. **29**(11), 2537–2551 (2017)

63. Yuan, J., Zheng, Y., Xie, X.: Discovering regions of different functions in a city using human mobility and POIs. In: Proceedings of the 18th ACM SIGKDD International Conference on Knowledge Discovery and Data Mining, pp. 186–194. ACM (2012)

64. Yuan, Q., Cong, G., Ma, Z., Sun, A., Thalmann, N.M.: Time-aware point-of-interest recommendation. In: Proceedings of the 36th International ACM SIGIR Conference on Research and Development in Information Retrieval, pp. 363–372. ACM (2013)

65. Yuan, Q., Cong, G., Sun, A.: Graph-based point-of-interest recommendation with geographical and temporal influences. In: Proceedings of the 23rd ACM International Conference on Conference on Information and Knowledge Management, pp. 659–668. ACM (2014)

66. Zhang, J., Kong, X., Yu, P.S.: Transferring heterogeneous links across location-based social networks. In: ACM International Conference on Web Search and Data Mining, pp. 303–312. ACM (2014)

67. Zhang, J.D., Chow, C.Y.: iGSLR: personalized geo-social location recommendation: a kernel density estimation approach. In: Proceedings of the 21st ACM SIGSPATIAL International Conference on Advances in Geographic Information Systems, pp. 334–343. ACM (2013)

68. Zhang, J.D., Chow, C.Y.: GeoSoCa: exploiting geographical, social and categorical correlations for point-of-interest recommendations. In: Proceedings of the 38th International ACM SIGIR Conference on Research and Development in Information Retrieval, pp. 443–452. ACM (2015)

69. Zhang, J.D., Chow, C.Y.: Ticrec: a probabilistic framework to utilize temporal influence corre-
    lations for time-aware location recommendations. IEEE Trans. Serv. Comput. **9**(4), 633–646
    (2016)
70. Zhang, J.D., Chow, C.Y., Li, Y.: LORE: exploiting sequential influence for location recommen-
    dations. In: Proceedings of the 22nd ACM SIGSPATIAL International Conference on Advances
    in Geographic Information Systems, pp. 103–112. ACM (2014)
71. Zhao, S., King, I., Lyu, M.R.: Capturing geographical influence in POI recommendations. In:
    International Conference on Neural Information Processing, pp. 530–537. Springer (2013)
72. Zhao, S., King, I., Lyu, M.R.: A survey of point-of-interest recommendation in location-based
    social networks (2016). arXiv:1607.00647
73. Zhao, S., King, I., Lyu, M.R.: Aggregated temporal tensor factorization model for point-of-
    interest recommendation. Neural Process. Lett. (2017). https://doi.org/10.1007/s11063-017-
    9681-8
74. Zhao, S., King, I., Lyu, M.R.: Geo-pairwise ranking matrix factorization model for point-of-
    interest recommendation. In: International Conference on Neural Information Processing, pp.
    368–377. Springer (2017)
75. Zhao, S., Lyu, M.R., King, I.: Aggregated temporal tensor factorization model for point-of-
    interest recommendation. In: International Conference on Neural Information Processing, pp.
    450–458. Springer (2016)
76. Zhao, S., Zhao, T., King, I., Lyu, M.R.: Geo-Teaser: Geo-Temporal sequential embedding rank
    for point-of-interest recommendation. In: Proceedings of the 26th International Conference
    on World Wide Web Companion, pp. 153–162. International World Wide Web Conferences
    Steering Committee (2017)
77. Zhao, S., Zhao, T., Yang, H., Lyu, M.R., King, I.: STELLAR: spatial-temporal latent ranking
    for successive point-of-interest recommendation. In: Thirtieth AAAI Conference on Artificial
    Intelligence, pp. 315–322 (2016)

Chapter 2
Understanding Human Mobility
from Geographical Perspective

Abstract POI recommendation is a significant service for LBSNs. It recommends
new places such as clubs, restaurants, and coffee bars to users. Whether recommended locations meet users' interests depends on three factors: user preference,
social influence, and geographical influence. Especially, capturing the geographical
influence plays the most important role for POI recommendations. Previous studies
observe that checked-in locations disperse around several centers and employ Gaussian distribution based models to approximate users' check-in behaviors. Yet centers
discovering methods are not satisfactory in prior work. This chapter shows how
to exploit Gaussian mixture model (GMM) and genetic algorithm based Gaussian
mixture model (GA-GMM) to capture geographical influence. Experimental results
on a real-world LBSN dataset show that GMM beats several popular geographical
capturing models in terms of POI recommendation, while GA-GMM excludes the
effect of outliers and enhances GMM.

Keywords POI Recommendation · Geographical influence
Gaussian mixture model

2.1 Introduction

POI recommendation is a significant service for LBSNs. With the development of
mobile devices and Web 2.0 technologies, many LBSNs like Foursquare and Gowalla
emerge and attract many users. These LBSNs allow users to check-in at POIs, make
friends, and share location-related information. In order to help users discover new
interesting places in LBSNs, the POI recommendation arises.

User preference, social influence, and geographical influence are three aspects
responsible for users' check-in activities [13, 14]. Generally we derive user preference from user-based collaborative filtering, explore social influence based on users'
social relationships, and model geographical influence from check-in locations' spatial features. And then we construct a POI recommendation system in the way of
combining those three kinds of influence. The representative work is as follows.

© The Author(s), under exclusive license to Springer Nature Singapore Pte Ltd.,
part of Springer Nature 2018
S. Zhao et al., *Point-of-Interest Recommendation in Location-Based Social Networks*,
SpringerBriefs in Computer Science, https://doi.org/10.1007/978-981-13-1349-3_2

Ye et al. [14] propose a linear fused framework to combine them and Cheng et al. [1] propose a fused model to recommend POIs.

For POI recommendation in LBSNs, research about geographical influence is new and requires more attention, comparing with user preference and social influence. It is well-defined on how to derive user preference and social influence in a recommendation system [6, 10]. Note that users' evaluations for items reflect their preferences and friends are inclined to share preferences. We derive user preference from user-based collaborative filtering and introduce social influence by containing similarity among friends. For POI recommendation system, we use collaborative filtering method to get user preference through treating locations as items and check-in frequencies as rating values, and we capture social influence by including friends' similarity in check-in locations [1, 13, 14]. In 2010, Ye et al. [13] first propose POI recommendation for LBSNs and utilize power law principle to model users' geographical influence. In the meantime, Cho et al. [2] study the user mobility in LBSNs inspired by Gonzalez's discovery [3]. The study focuses on those users who frequently check-in, since Gonzalez's discovery bases on call logs data that have strong periodic property. They propose a periodic mobility model (PMM) to capture user's geographical influence for location prediction in LBSNs. Next, Cheng et al. [1] propose Multi-center Gaussian Model (MGM) to capture geographical influence. This model assumes a user's visited locations disperse around several centers and utilizes a greedy method to discover centers. It defines a district by a fixed distance and thus ignores discrepancy between users. In summary, Gaussian distribution based models perform well for POI recommendation, but we still encounter challenges in how to discover the activity centers accurately and how to eliminate the effect of outliers.

In order to find activity centers more accurately and eliminate outliers, we propose two models—Gaussian mixture model (GMM) and genetic algorithm based Gaussian mixture model (GA-GMM) to capture geographical influence. From geographical perspective, people prefer places that are nearer to their activity centers. Those frequent checked-in places naturally form a user's activity district. According to locations' spatial clustering feature, we apply GMM to find a user's activity district centers. However, outliers exist in the observed data that do harm to learn the model. How to eliminate the impact of outliers? Thang et al. [11] propose a genetic algorithm based EM algorithm to implement the trimmed likelihood estimate (TLE) method [9] to eliminate the outliers in mixture models. We exploit this genetic based EM algorithm to train GMM. The genetic algorithm based GMM (GA-GMM) improves GMM and models the geographical influence better.

This chapter bases on the published work [15] with the contributions as follows. First, we propose GMM to automatically learn users' activity centers via exploring their check-in history records. Moreover, we enhance GMM by GA-GMM to eliminate outliers. Finally, we conduct experiments on a real-world LBSN dataset and demonstrate that the proposed models capture the geographical information better and improve the accuracy of POI recommendation.

2.2 Related Work

In this section, we introduce related work in three aspects: POI recommendation in LBSNs, geographical influence capturing methods, and GA-GMM.

POI recommendation in LBSNs is a new research topic. POI recommendation is widely used in GPS-based mobile devices at first [4, 5]. In 2010, Ye et al. [13] first propose POI recommendation in LBSNs. Further, Ye et al. [14] point out that user preference, social influence, and geographical influence are three aspects responsible for recommending POIs and geographical influence is the most important among the three factors. The representative work is as follows. Ye et al. [14] recommend POIs through a linear fused framework combining user preference, social influence, and geographical influence. Cheng et al. [1] propose a fused model to combine them to recommend POIs.

Study of geographical influence capturing methods is new for POI recommendation. In 2010, Ye et al. [13] first propose POI recommendation for LBSNs and arise a power law principle to capture geographical influence for POI recommendation. Earlier related work about geographical influence appears in the study of user movement pattern. Gonzalez et al. [3] build a model using call logs and discover that activities of an individual usually center around a small number of frequently visited locations. Based on this, Cho et al. [2] study the specific users frequently checking in and propose a periodic mobility model (PMM) to capture geographical influence for location prediction in LBSNs. Cheng et al. [1] employ a multi-center Gaussian model (MGM) to capture the geographical feature of locations in the proposed fused POI recommendation model.

Genetic algorithm based GMM (GA-GMM) is a method to eliminate outliers when learning GMM. Trimmed likelihood estimate (TLE) method is adopted to eliminate outliers in some studies of mixture model analysis [9]. Thang et al. [11] first propose a genetic algorithm based method to implement the trimmed likelihood estimating method to train mixture models and demonstrate the performance through a genetic algorithm based GMM (GA-GMM). Wang et al. utilize the GA-GMM to process EEG signal and apply it on brain-computer interface [12].

2.3 Model

2.3.1 Gaussian Mixture Model

Gaussian mixture model (GMM) [8] is the most widely used mixture model. We can formulize it as follows:

$$p(x_i) = \sum_{k=1}^{K} \pi_k \mathcal{N}(x_i | \mu_k, \sum\nolimits_k),$$

where $p(x_i)$ denotes probability dense distribution of data x_i, μ_k indicates mean value, $\sum_k$ indicates covariance matrix for a base distribution, K denotes the number of base components, and π_k is the mixing coefficient.

We exploit GMM to capture geographical influence in POI recommendation. Each Gaussian distribution component represents an activity district and the mean value denotes the longitude and latitude of the district center. Centers may be his home, office, or some specific entertainment place. We assume places nearer to some center are geographically easier to arrive and people prefer those places.

In the following, we show how to recommend POIs through GMM. For a user, a location's geographical information ([longitude, latitude]) in his check-in history records represents data x_i. We recommend POIs through the following steps:

1. Learn the parameters of GMM,
2. Calculate candidate locations' probabilities fitting the trained model, and
3. Sort the candidate locations and recommend the top K locations.

2.3.2 Genetic Algorithm Based Gaussian Mixture Model

In order to eliminate the effect of outliers, we introduce a genetic algorithm based Gaussian mixture model (GA-GMM). Generally we could use maximum likelihood EM algorithm to learn GMM [8]. If we use Θ to denote the parameters, likelihood function could be represented as

$$p(X|\Theta)_{ML} = \prod_{i=1}^{n} p(x_i|\Theta).$$

Further, if we use the logarithm form, we can denote the objective of maximum likelihood EM algorithm as follows:

$$\hat{\Theta}_{ML} = \arg\max \log p(X|\Theta)_{ML} = \arg\max \sum_{i=1}^{n} \log p(X_i|\Theta). \qquad (2.1)$$

This formula includes all observed data. Trimmed likelihood estimate (TLE)—that aims to to select the subset of data with maximum sum of likelihood values—is used to eliminate the outliers [9]. We can use a genetic algorithm to find the optimal subset and exploit maximum likelihood EM algorithm to learn the parameters of GMM, as illustrated in Algorithm 2 [11]. In this case, the objective function could be represented as

$$\log p_{TLE}(X|\Theta) = \sum_{i=1}^{n} w_i \log p(X_i|\Theta), \qquad (2.2)$$

where $\forall i = 1, 2, \ldots, n$, $w_i \in \{0, 1\}$ and $\sum_{i=1}^{n} w_i = m$, m represents the number of valid data. When $w_i = 1$, it indicates that the corresponding data is chosen into the subset. Otherwise, the data is an outlier and should be discarded. Hence, the result is a subset of size m out of n original samples, which fits GMM most in terms of likelihood contribution.

As a genetic algorithm, GA-GMM contains properties of genetic algorithm—it includes encoding scheme, fitness function, and operators like crossover, mutation, and selection. We use the standard way to implement crossover and selection [7]. Encoding scheme, fitness function, and a self-defined mutation (Guided Mutation) are defined as follows.

Definition 2.1 Encoding scheme. The chromosome is encoded into a binary string and each bit represents the existence of corresponding observed data. Each chromosome and its corresponding mixture model will be a possible solution to our problem.

Definition 2.2 Fitness function. The fitness score function is set as the trimmed logarithm likelihood of the corresponding GMM of a chromosome—$\log p_{TLE}(X|\Theta)$.

Definition 2.3 Guided Mutation. Guided Mutation ensures the chromosome in a population to mutate toward maximizing fitness score. It means we choose chromosome that has higher value fitting trained GMM.

Algorithm 2: Genetic-based Expectation Maximization Algorithm

---

1. t=0;
2. Initialize $P_0(t)$;
3. **for** $t = 1 : G$ **do**
4. $P_1(t) \leftarrow$ perform several cycles of EM on $P_0(t)$;
5. $P_2(t) \leftarrow$ Guided Mutation in $P_1(t)$;
6. $fScore_2 \leftarrow$ evaluate $P_2(t)$;
7. $P_0(t)' \leftarrow$ selection and crossover to generate offspring from $P_2(t)$;
8. $P_1(t)' \leftarrow$ perform several cycles of EM on $P_0(t)'$;
9. $P_2(t)' \leftarrow$ Guided Mutation in $P_1(t)'$;
10. $fScore_2' \leftarrow$ evaluate $P_2(t)'$;
11. $P_3(t) \leftarrow$ selection from $[P_2(t), P_2(t)']$;
12. $iBest \leftarrow$ best individual from $P_3(t)$;
13.   **if** $iBest$ satisfies convergence condition **then** break;
14. $P_0(t+1) \leftarrow P_3(t)$;
15. $t = t + 1$;
16. Perform EM on $iBest$ until convergence;

Table 2.1 Data statistics

Min. C.	Max. C.	Avg. C.	Min. T.	Max. T.	Avg. T.
1,001	50,243	2,505	366	968	593

2.4 Experiment

2.4.1 Setup and Metrics

We prepare the data by cleaning and splitting. We filter locations of less than 10 visits. And then we split the dataset into three non-overlapping sets in sequence: a redundant set, a training set, and a test set. The test set keeps 10% of the whole data set. We test different cases in which the proportion of training data is 90% and 50% respectively. When training data set is 90%, there is no redundant data. When the training data set is 50%, redundant data is the former 40% data that will be discarded.

We evaluate the performance of different models in capturing geographical influence by the accuracy of POI recommendation that is measured by Precision and Recall. POI recommendation is to recommend the top-N highest ranked locations. However, the system should not recommend locations the user has checked in. To evaluate the performance of POI recommendation, we use the Precision@N and Recall@N as the metrics that are standard metrics to measure the performance of POI recommendation [14]. Precision@N defines the ratio of recovered POIs to the N recommended POIs and Recall@N defines the ratio of recovered POIs to the size of test set.

2.4.2 Dataset

We use the Gowalla data records from February 2009 to September 2011. We select 3836 active users' records to experiment. We define active users as users whose check-ins are more than 1000 times and experience of using Gowalla is more than 1 year. After removing locations with less than 10 visits, all check-ins of active users include 183,667 different locations. We illustrate statistics of the data in Table 2.1, where "C." represents the check-in times of a user and "T." represents the time span (unit is day) from first check-in to last check-in.

2.4.3 Results

We compare the POI recommendation performance of GMM and GA-GMM with Gaussian model (GM) and Multi-center of Gaussian model (MGM) [1] when training data set is 90% and 50% respectively.

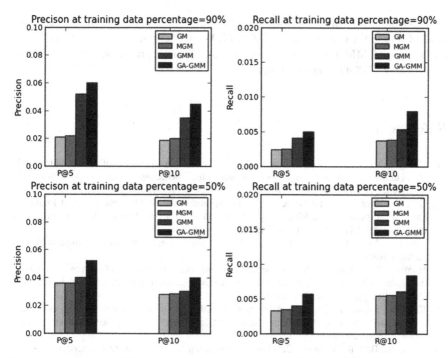

Fig. 2.1 Comparison of different models [15]

Gaussian model (GM) [3] is a baseline model used in [2]. It models human movement as a stochastic process centered around a single point.

Multi-center Gaussian model (MGM) [1] is a latest model. It uses a fixed distance to define a district. When check-ins in a district are more than a threshold, the mean of all check-ins is the center. It utilizes a greedy method to find the district and requires no overlapping between two districts.

We illustrate experimental results in Fig. 2.1. GMM outperforms GM and MGM; further GA-GMM improves GMM. Hence, GA-GMM could better capture the geographical influence. In the experiment, we set the number of centers in GMM and GA-GMM as 2 for simplicity, since Cho et al. propose that the check-in behavior comprises two states in [2]. We set the radius of a region in MGM as 1 km and the threshold as 10% (that means the ratio of check-ins in one district is at least 10% of all his check-ins).

2.5 Conclusion

We apply GMM and GA-GMM to capture geographical influence in POI recommendation. We exploit GMM to automatically learn users' activity centers; further we utilize GA-GMM to improve GMM by eliminating outliers. According to experimental results, we draw conclusions as follows. (1) GMM outperforms the baseline model GM and the latest model MGM. (2) GA-GMM eliminates the outliers of data and improves GMM. It discovers the activity centers more precisely, which increases the accuracy of POI recommendation.

References

1. Cheng, C., Yang, H., King, I., Lyu, M.R.: Fused matrix factorization with geographical and social influence in location-based social networks. In: Proceedings of the Twenty-Sixth AAAI Conference on Artificial Intelligence, pp. 17–23. AAAI Press (2012)
2. Cho, E., Myers, S.A., Leskovec, J.: Friendship and mobility: user movement in location-based social networks. In: Proceedings of the 17th ACM SIGKDD International Conference on Knowledge Discovery and Data Mining, pp. 1082–1090. ACM (2011)
3. Gonzalez, M.C., Hidalgo, C.A., Barabasi, A.L.: Understanding individual human mobility patterns. Nature **453**(7196), 779–782 (2008)
4. Horozov, T., Narasimhan, N., Vasudevan, V.: Using location for personalized poi recommendations in mobile environments. In: International Symposium on Applications and the Internet (SAINT'06), pp. 6–11. IEEE (2006)
5. Kang, E.y., Kim, H., Cho, J.: Personalization method for tourist point of interest (poi) recommendation. In: Knowledge-Based Intelligent Information and Engineering Systems, pp. 392–400. Springer (2006)
6. Koren, Y., Bell, R., Volinsky, C.: Matrix factorization techniques for recommender systems. Computer **42**(8), 30–37 (2009)
7. Melanie, M.: An introduction to genetic algorithms. Cambridge, Massachusetts London, England, Fifth printing **3** (1999)
8. Murphy, K.P.: Machine learning: a probabilistic perspective. The MIT Press (2012)
9. Neykov, N., Filzmoser, P., Dimova, R., Neytchev, P.: Robust fitting of mixtures using the trimmed likelihood estimator. Comput. Stat. Data Anal. **52**(1), 299–308 (2007)
10. Ricci, F., Shapira, B.: Recommender Systems Handbook. Springer (2011)
11. Thang, N.D., Lihui, C., Keong, C.C.: An outlier-aware data clustering algorithm in mixture models. In: 7th International Conference on Information, Communications and Signal Processing, 2009. ICICS 2009, pp. 1–5. IEEE (2009)
12. Wang, B., Wong, C.M., Wan, F., Mak, P.U., Mak, P.I., Vai, M.I.: Gaussian mixture model based on genetic algorithm for brain-computer interface. In: Image and Signal Processing (CISP), 2010 3rd International Congress on, vol. 9, pp. 4079–4083. IEEE (2010)
13. Ye, M., Yin, P., Lee, W.C.: Location recommendation for location-based social networks. In: Proceedings of the 18th SIGSPATIAL International Conference on Advances in Geographic Information Systems, pp. 458–461. ACM (2010)
14. Ye, M., Yin, P., Lee, W.C., Lee, D.L.: Exploiting geographical influence for collaborative point-of-interest recommendation. In: Proceedings of the 34th International ACM SIGIR Conference on Research and Development in Information Retrieval, pp. 325–334. ACM (2011)
15. Zhao, S., King, I., Lyu, M.R.: Capturing geographical influence in poi recommendations. In: International Conference on Neural Information Processing, pp. 530–537. Springer (2013)

Chapter 3
Understanding Human Mobility from Temporal Perspective

Abstract Understanding user mobility from the temporal perspective is the key to POI recommendation that mines user check-in sequences to suggest interesting locations for users. Because user mobility in LBSNs exhibits strong temporal patterns—for instance, users would like to check-in at restaurants at noon and visit bars at night. Hence, capturing the temporal influence is necessary to ensure the high performance in a POI recommendation system. This chapter summarizes the temporal characteristics of user mobility in LBSNs in three aspects: periodicity, consecutiveness, and non-uniformness. Moreover, an Aggregated Temporal Tensor Factorization (ATTF) model for POI recommendation is proposed to capture the three temporal features. Experiments on two real-world datasets show that the ATTF model achieves better performance than the state-of-the-art temporal models for POI recommendation.

Keywords POI recommendation · Temporal influence · Tensor factorization

3.1 Introduction

Understanding the user mobility from temporal perspective is important to establish a practical POI recommendation system. Previous studies show that the user mobility in LBSNs exhibits significant temporal features [3, 4, 32]. For example, users always stay in the office in the Monday afternoon, and enjoy entertainments in bars at night. In summary, the temporal features in users' check-in data can be abstracted in three aspects.

- **Periodicity**. Users share the same periodic pattern, visiting the same or similar POIs at the same time slot [3, 32]. For instance, a user always visits restaurants at noon, so do other users. Hence, the periodicity inspires the time-aware collaborative filtering method to recommend POIs [32].
- **Consecutiveness**. A user's current check-in is largely correlated with the recent check-in [2, 4]. Gao et al. [4] model this property by assuming that user preferences are similar in two consecutive hours. Cheng et al. [2] assume that two checked-in POIs in a short term are highly correlated in latent feature space.

© The Author(s), under exclusive license to Springer Nature Singapore Pte Ltd., part of Springer Nature 2018
S. Zhao et al., *Point-of-Interest Recommendation in Location-Based Social Networks*, SpringerBriefs in Computer Science, https://doi.org/10.1007/978-981-13-1349-3_3

- **Non-uniformness**. A user's check-in preference changes at different hours of a day [4]. For example, at noon a user may visit restaurants while at night the user may have fun in bars.

By capturing the observed temporal features, a variety of systems are proposed to enhance POI recommendation performance [2, 4, 32], which gain better performance than general collaborative filtering (CF) methods [30]. Nevertheless, previous work [2, 4, 32] cannot model the three features together. Moreover, an important fact is ignored in prior work that the temporal influence exists at different time scales. For example, in day level, you may check-in at POIs around your home in the earning morning, visit places around your office in the day time, and have fun at nightclubs in the evening. In week level, you may stay in the city for work on weekdays and go out for vocation on weekends. Hence, to better model the temporal influence, capturing the temporal features at different time scales is necessary.

This chapter proposes an Aggregated Temporal Tensor Factorization (ATTF) model for POI recommendation to capture the three temporal features together, as well as at different time scales. We construct a user-time-POI tensor to represent the check-ins as shown in Fig. 3.1, and then employ the interaction tensor factorization [26] to model the temporal effect. Different from prior work that represents the temporal influence at a single scale, we index the temporal information at different scales, i.e., hour, week day, and month, to learn the latent representation. Furthermore, we employ a linear combination operator to aggregate different temporal latent features' contributions, which capture the temporal influence at different scales. Specifically, the ATTF model learns the three temporal properties as follows: (1) periodicity is learned from the temporal CF mechanism; (2) consecutiveness is manifested in two aspects—time in a slot brings the same effect through sharing the same time factor, and the relation between two consecutive time slots can be learned from the tensor model; (3) non-uniformness is depicted by different time factors representing different time slots from each time scale perspective. Moreover, an aggregate operator is introduced to combine the temporal influence at different scales, i.e., hour, week day, and month, and represent the temporal effect in a whole. Moreover, we establish an embedding neural network to represent the ATTF model,

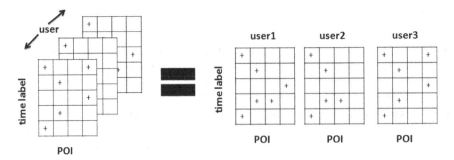

Fig. 3.1 Tensor illustration for check-ins

which gives new insights to understand the proposed model from the neural network perspective.

To sum up, this chapter bases on the published work [36] with the following contributions.

- To the best of our knowledge, this is the first temporal tensor factorization method for POI recommendation, subsuming all the three temporal properties: periodicity, consecutiveness, and non-uniformness.
- We propose a novel model to capture temporal effect in POI recommendation at different time scales. Experimental results show that our model outperforms prior temporal model more than 20%.
- The ATTF model is a general framework to capture the temporal features at different scales, which outperforms single temporal factor model and gains 10% improvement in the top-5 POI recommendation task on Gowalla data.
- We understand the ATTF model from the embedding neural network perspective, verifying the effectiveness of the embedding neural network that is a general framework for latent factor models, including rating estimation models (e.g., MF [9]) and ranking models (e.g., our ATTF model).

3.2 Related Work

In this section, we first review the literature of POI recommendation. Then, we summarize the progress of modeling temporal effect for POI recommendation. Finally, we review the literature of embedding learning and its applications, which inspires us to understand the proposed ATTF model from the neural network perspective.

POI Recommendation. Most of POI recommendation systems base on the collaborative filtering (CF) techniques, which can be reported in two aspects, memory-based and model-based. On the one hand, Ye et al. [30] propose the POI recommendation problem in LBSNs solved by user-based CF method, and further improve the system by linearly combining the geographical influence, social influence, and preference similarity. In order to enhance the performance, more advanced techniques are then applied, e.g., incorporating temporal influence [32], and utilizing a personalized geographical model via kernel density estimator [33, 34]. On the other hand, model-based CF is proposed to tackle the POI recommendation problem that benefits from its scalability. Cheng et al. [1] propose a multi-center Gaussian model to capture user geographical influence and combine it with social matrix factorization (MF) model [19] to recommend POIs. Gao et al. [4] propose an MF-based model, Location Recommendation framework with Temporal effects (LRT), utilizing similarity between time-adjacent check-ins to improve performance. Lian et al. [14] and Liu et al. [18] enhance the POI recommendation by incorporating geographical information in a weighted regularized matrix factorization model [8]. In addition, some researchers subsume users' comments to improve the recommendation performance [5, 13, 31]. Other researchers model the consecutive check-ins' correlations to enhance the system [2, 16, 37, 38].

Temporal Effect Modeling. In 2011, Cho et al. [3] propose the periodicity of check-in data in LBSNs. People always visit restaurants at noon, so we suffice to recommend users restaurants meeting their tastes at noon. The CF technique helps us to recommend similar POIs at the same time slot. However, experiments in [3] depend on dense check-in data, not fitting most of the users. In 2013, Yuan et al. [32] combine the temporal similarity and non-temporal similarity and propose a new similarity metric to enhance the user-based CF model. At the same year, Gao et al. [4] observe the non-uniformness property (a user's check-in preference changes at different hours of a day), and consecutiveness (a user's preference at time t is similar with time $t - 1$). Further, Gao et al. propose LRT model based on MF technique to model the non-uniformness and consecutiveness. Meantime, Cheng et al. [2] propose the Factorized Personalized Markov Chain model [25] with Local Region constraint (FPMC-LR) to capture the consecutiveness, supposing the strong correlation between two consecutive checked-in POIs. However, previous work does not model the three features together nor modeling the temporal influence at different scales.

Embedding Neural Network. The embedding neural network, e.g., word2vec framework [20], has turned out to be a successful semi-supervised learning method. It is used in natural language processing [15, 17]. For the efficacy of the framework in capturing the correlations of items, the embedding neural network is employed to the network embedding [7, 23, 28], and as well as in recommendation systems [27, 29]. Moreover, recent studies [11, 12] show that the neural word embeddings can be treated as a kind of matrix factorization method [9]. This equivalence between neural embeddings and the latent factor models inspires us to understand our ATTF model from the embedding neural network perspective. Our interpretation of ATTF model from neural network perspective verifies that the embedding neural network can be treated as a general framework for latent factor models, including rating estimation models [9] and ranking models (e.g., our proposed ATTF model).

3.3 Preliminaries

In this section, we first analyze the temporal features of user check-in data. Then, we introduce the time labeling scheme that is the prerequisite of our ATTF model. We analyze user check-in data in Foursquare and Gowalla, which demonstrates the similar check-in pattern. In the following, we show the empirical data analysis result based on a randomly selected user in Foursquare.

3.3.1 Empirical Data Analysis

We leverage empirical data analysis to explore the three temporal properties of check-in data. Our analysis verifies previous discoveries, for instance, the non-uniformness—user check-in preference changes at different time of a day [4].

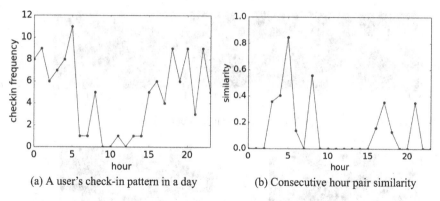

Fig. 3.2 Sparsity demonstration [36]

Moreover, we observe some new properties not covered in prior work, e.g., the non-uniformness exists at different time scales.

Data sparsity is a big concern in previous temporal models. Figure 3.2a demonstrates a user's check-in pattern in a day. We observe that the user always has many check-ins in the morning and evening, which verifies the periodicity. The check-in activity repeats in the morning and evening. Figure 3.2b shows the consecutive hour pair similarity,[1] i.e., the check-in similarity between time t and $t - 1$ (t means the hour $1, 2, \ldots, 24$). We observe that the user check-in preference has high similarity at some consecutive hours, e.g., between 5 o'clock and 4 o'clock, 8 o'clock and 7 o'clock. However, we also find that at some time (e.g., 9:00), the user has few check-ins and the similarity is zero. Therefore, the sparse data make it too hard to model the periodicity and consecutiveness via a counting method (e.g., Pearson correlation or cosine similarity). As shown in Fig. 3.2b, most of the similarities are zero. So the consecutiveness cannot be modeled at most of the time. The dilemma of counting methods in the face of sparse data motivates us to exploit a latent factor learning model. In our model, we use a time latent factor to represent the temporal effect of a time slot, not modeling the temporal effect from the user or POI perspective. Further, the temporal factor is learned from all users' check-ins at the time slot. Therefore, it overcomes the sparsity problem in counting methods.

We observe that the non-uniformness (e.g., the check-in change characteristics) exists at different time scales in users' check-in data. Following [4], we demonstrate an example of a random user's aggregated check-in activities on his/her top 5 most visited POIs in Fig. 3.3. Figure 3.3a verifies the non-uniformness: a user's check-in preference changes at different hours of a day [4]. As shown in Fig. 3.3a, the most visited POI changes at different hours. For example, the most visited POI is POI 1 at 1:00 while the most visited POI is POI 4 at 5:00. Besides, we discover there are other change characteristics. As shown in Fig. 3.3b, c a user's check-in preference changes at different months of a year, and among different days of a week as well.

---

[1]We use cosine similarity here; other measures like Pearson correlation are also applicable.

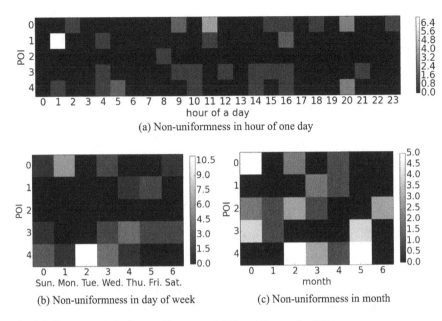

(a) Non-uniformness in hour of one day

(b) Non-uniformness in day of week

(c) Non-uniformness in month

Fig. 3.3 Demonstration of non-uniformness at different time scales [36]

The change of check-ins at different time scales depicts the user preference from different perspectives: (1) A user may check-in at POIs around his/her home in the morning, visit places around the office in the day time, and have fun in bars at night. (2) A user may visit more locations around his/her home or office on weekdays. On weekends, he/she may check-in more at some shopping malls or vocation places. (3) At different months, a user may have different customs. For instance, he/she would visit ice cream shops in the months of summer and hot pot restaurants in the months of winter. Hence, only modeling the non-uniformness at a single scale, we cannot capture all temporal features, which need to be formulated at different scales.

3.3.2 Time Labeling Scheme

Time labeling is a prerequisite of our ATTF model. We use a time latent factor to represent the temporal effect at a specific time, and then learn from a latent factor model. Time labeling scheme determines how to assign a latent factor to specific time. Before diving to the model, we describe the time labeling scheme first.

Figure 3.4 demonstrates the time labeling scheme. In order to capture temporal features at different time scales, we represent a time spot with several parts and then aggregate their contributions together. According to the empirical data analysis, we consider temporal features in three time scales: month of a year, day of a week, and hour of a day. Now the temporal effect is formulated by three latent time factors. As

Fig. 3.4 Time labeling
scheme demonstration [36]

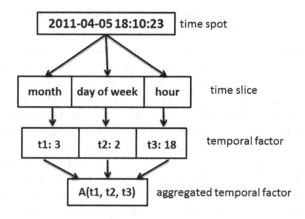

shown in Fig. 3.4, we leverage three slices to denote a time spot: month of year, day of week, and hour of day. Further they are depicted by three kinds of different temporal latent vectors respectively. So a time spot t is labeled by a tuple (t_1, t_2, t_3), which satisfies that $t_1 \in \{0, 1, \ldots, 12\}$, $t_2 \in \{0, 1, \ldots, 6\}$, $t_3 \in \{0, 1, \ldots, 23\}$ (we have 12 months in a year, 7 days in a week, and 24 hours in a day). Furthermore, we define $T_1 \subset R^{12 \times d}$, $T_2 \subset R^{7 \times d}$, and $T_3 \subset R^{24 \times d}$ to denote the corresponding temporal latent factor matrices, where d is the latent vector dimension.

To aggregate several temporal factors, we define an operator $A(\cdot) : R^d \times R^d \times R^d \rightarrow R^d$ to combine different temporal features. Take "2011-04-05 18:10:23" as an example, its label ids at month, day of week, and hour are 3 (April), 2 (Tuesday), 18 (after 18:00). Hence its temporal latent factor is formulated as $A(T_{1,3}, T_{2,2}, T_{3,18})$. It is important to note that our scheme is flexible: we are able to ignore one feature by taking away a slice, or introduce a new feature by adding a new slice.

Memory Reducing Trick. We reduce the input data size through a binary coding trick. We employ one label id instead of three to represent the three slices. In detail, we use 4 bits to represent the month, 3 bits to represent the day of week, and 5 bits to represent the hour slot. So the time label id can be represented by an integer of 16 bits. For instance, "2011-04-05 18:10:23" could be coded as "0011 010 10010", and its label id is 850. When we learn the model, we transform the label id into binary representation and find the corresponding label to each slice. After labeling the time, we are able to model the temporal effect from a user-time-POI latent factorization model; see details in Sect. 3.4.

3.4 Method

In this section, we first demonstrate the ATTF model. Then we give the detailed model inference and learning procedure. Finally, we summarize the model discussion.

3.4.1 Aggregated Temporal Tensor Factorization Model

Denote that $\mathcal{U}$ is the set of users and $\mathcal{L}$ is the set of POIs. In addition, $\mathcal{T}_1$, $\mathcal{T}_2$, and $\mathcal{T}_3$ are the set of months, days of week, and hours respectively. Further we define $\mathcal{T}$ as the set of time label tuples, consisting of elements $t := (t_1, t_2, t_3)$, namely the temporal representation at different scales. The ATTF model estimates the preference of a user u at a POI l given a specific time label t through a score function $f(u, t, l)$, where $u \in \mathcal{U}$ is user id, $t \in \mathcal{T}$ is a time label tuple, and $l \in \mathcal{L}$ is POI id.

We are typically given N training examples $(u_i, t_i, l_i) \in \{1, \ldots, |\mathcal{U}|\} \times \{1, \ldots, |\mathcal{T}|\} \times \{1, \ldots, |\mathcal{L}|\}, i = 1, 2, \ldots, N$, and correspondingly outputs $y_i \in R$, $i = 1, 2, \ldots, N$. Here, (u_i, t_i, l_i) is the index of a particular element of a user-time-POI tensor, and y_i is the preference score of the user at the POI given the time label. One could simply collate the training data to build a suitable tensor, so the training task turns to fill in the blank entries of the tensor.

We exploit the Pairwise Interaction Tensor Factorization (PITF) [26] model to decompose the user-time-POI tensor. PITF model that turns out to be successful in the ECML/PKDD Discovery Challenge, runs much faster than other tensor factorization methods and has better performance in a large scale prediction task [26]. Thus the score of a POI l given user u and time t is factorized into three interactions: user-time, user-POI, and time-POI, where each interaction is modeled through the latent vector product. Further, we infer the model via Bayesian Personalized Ranking (BPR) criteria [24] that is a general framework to train a recommendation system from implicit feedback. Because prior work [14, 37, 38] indicates that treating the check-ins as implicit feedback is better than explicit ways for POI recommendation. Since we recommend POIs for users at specific time, any candidate POI has the same user-time interaction. As a result, the preference score is independent of the user-time interaction. Then the score function for a given time label t, user u, and a target POI l could be formulated as:

$$f(u, t, l) = \langle U_u^{(L)}, L_l^{(U)} \rangle + \langle A(T_{1,t_1}^{(L)}, T_{2,t_2}^{(L)}, T_{3,t_3}^{(L)}), L_l^{(T)} \rangle, \tag{3.1}$$

where $\langle \cdot \rangle$ denotes the vector inner product, $A(\cdot)$ is the aggregate operator. Suppose that d is the latent vector dimension, $U_u^{(L)} \in R^d$ is user u's latent vector for POI interaction, $L_l^{(U)}, L_l^{(T)} \in R^d$ are POI l's latent vectors for user interaction and time interaction, $T_{1,t_1}^{(L)}, T_{2,t_2}^{(L)}, T_{3,t_3}^{(L)} \in R^d$ are time t's latent vector representations in three aspects: month, day of week, and hour.

Aggregate operator combines the several temporal features together. We propose a linear convex combination operator. It is formulated as follows,

$$A(\cdot) = \alpha_1 \cdot T_{1,t_1}^{(L)} + \alpha_2 \cdot T_{2,t_2}^{(L)} + \alpha_3 \cdot T_{3,t_3}^{(L)}, \tag{3.2}$$

where α_1, α_2, and α_3 denote the weights of each temporal factor, which satisfy $\alpha_1 + \alpha_2 + \alpha_3 = 1$, and $\alpha_1, \alpha_2, \alpha_3 >= 0$.

3.4.2 Learning

We infer the model via BPR criteria [24], which treats the check-in activity as a kind of implicit feedback. Namely, we assume the user prefers the visited POIs than the unvisited. We treat the visited POIs as positive and the unvisited as negative. Then, we suppose that the score of $f(u, t, l)$ at positive observations is higher than the negative POIs, given u and t. Further, we formulate the relation that user u prefers a positive POI l_i than a negative one l_j at time t as follows

$$l_i >_{u,t} l_j. \tag{3.3}$$

Based on the pairwise preference defined above, we suffice to extract the set of preference constraints from the training examples

$$D_S := \{(u, t, l_i, l_j) | l_i >_{u,t} l_j, u \in \mathcal{U}, t \in \mathcal{T}, l_i, l_j \in \mathcal{L}\}. \tag{3.4}$$

For simplicity, we denote $y_{u,t,l} = f(u, t, l)$. Then for any quadruple in D_S, it satisfies $y_{u,t,l_i} > y_{u,t,l_j}$. Using a logistic function to model this relation, we get

$$p(l_i >_{u,t} l_j) := \sigma(y_{u,t,l_i} - y_{u,t,l_j}), \tag{3.5}$$

which measures the probability of l_i is a positive observation and l_j is a negative observation for user u at time t. In Eq. (3.5), σ is the logistic function $\sigma(x) = \frac{1}{1+e^{-x}}$.

Suppose the quadruples in D_S are independent of each other, then learning the ATTF model is to maximize the likelihood of all the pair orders

$$\arg \max_{\Theta} \prod_{(u,t,l_i,l_j) \in D_S} p(l_i >_{u,t} l_j), \tag{3.6}$$

where Θ is the parameters to learn, namely $U^{(L)}, L^{(U)}, L^{(T)}, T_1^{(L)}, T_2^{(L)}$, and $T_3^{(L)}$. The objective function is equivalent to minimizing the negative log likelihood. To avoid the risk of overfitting, we add a Frobenius norm term to regularize the parameters. Then the objective function is

$$\arg \min_{\Theta} \sum_{(u,t,l_i,l_j) \in D_S} - ln(\sigma(y_{u,t,l_i} - y_{u,t,l_j})) + \lambda_{\Theta} ||\Theta||_F^2, \tag{3.7}$$

where λ_{Θ} is the regularization parameter.

We leverage the Stochastic Gradient Decent (SGD) algorithm to learn the objective function for efficacy. First, we define $y_{u,t,l_p,l_n} = y_{u,t,l_p} - y_{u,t,l_n}$, which models the pairwise relation in D_S. Further we denote a common part in gradient decent values for all parameters as $\delta = 1 - \sigma(y_{u,t,l_p,l_n})$. As $T_{1,t_1}^{(L)}, T_{2,t_2}^{(L)}$, and $T_{3,t_3}^{(L)}$ are symmetric, they have the same gradient form. For simplicity, we use $T_t^{(L)} \in \{T_{1,t_1}^{(L)}, T_{2,t_2}^{(L)}, T_{3,t_3}^{(L)}\}$

to represent any of them, $\alpha \in \{\alpha_1, \alpha_2, \alpha_3\}$ to denote corresponding weight, and $A(\cdot)$ to denote $A(T_{1,t_1}^{(L)}, T_{2,t_2}^{(L)}, T_{3,t_3}^{(L)})$. Then the updating rule for the parameters is as follows,

$$
\begin{aligned}
U_u^{(L)} &\leftarrow U_u^{(L)} + \gamma \cdot (\delta \cdot (L_{l_p}^{(U)} - L_{l_n}^{(U)}) - \lambda \cdot U_u^{(L)}), \\
L_{l_p}^{(U)} &\leftarrow L_{l_p}^{(U)} + \gamma \cdot (\delta \cdot U_u^{(L)} - \lambda \cdot L_p^{(U)}), \\
L_{l_p}^{(T)} &\leftarrow L_{l_p}^{(T)} + \gamma \cdot (\delta \cdot A(\cdot) - \lambda \cdot L_{l_p}^{(T)}), \\
L_{l_n}^{(U)} &\leftarrow L_{l_n}^{(U)} - \gamma \cdot (\delta \cdot U_u^{(L)} + \lambda \cdot L_n^{(U)}), \\
L_{l_n}^{(T)} &\leftarrow L_{l_n}^{(T)} - \gamma \cdot (\delta \cdot A(\cdot) + \lambda \cdot L_{l_n}^{(T)}), \\
T_t^{(L)} &\leftarrow T_t^{(L)} + \gamma \cdot (\delta \cdot \alpha \cdot (L_{l_p}^{(T)} - L_{l_n}^{(T)}) - \lambda \cdot T_t^{(L)}),
\end{aligned}
\tag{3.8}
$$

where γ is the learning rate, λ is the regularization parameter. To train the model, we use the bootstrap skill to draw the quadruple from D_S, following [24]. Algorithm 3 gives the detailed procedure to learn the ATTF model. We aim to learn the latent representations of user, temporal features, and POIs, namely $U^{(L)}$, $T_1^{(L)}, T_2^{(L)}, T_3^{(L)}$, $L^{(U)}, L^{(T)}$. Let $|\mathcal{U}|$ denote the number of users, then we generate about $\lfloor 100 * \sqrt{|\mathcal{U}|} \rfloor$ tuples from D_S to generate a tuple set D_e for the loss estimation, namely the negative log likelihood value. We follow the implementation of BPRMF [24] in MyMediaLite[2] to set the number of samples for loss estimation as $\lfloor 100 * \sqrt{|\mathcal{U}|} \rfloor$. In each iteration, we sample S check-ins and then generate negative samples to learn the model. After that, we calculate the loss value over D_e: $\sum_{(u,t,l_i,l_j) \in D_e} -ln(\sigma(y_{u,t,l_i} - y_{u,t,l_j})) + \lambda_\Theta ||\Theta||_F^2$. The convergent condition is satisfied when the loss value for the fixed sampled tuples does not decrease.

Complexity. The runtime for predicting a triple (u, t, l) is in $O(d)$, where d is the number of latent vector dimension. The updating procedure is also in $O(d)$. Hence training a quadruple is in $O(d)$, then training an example (u, t, l) is in $O(k \cdot d)$, where k is the number of sampled negative POIs. For each iteration, we sample S training examples. The calculation cost for loss estimation is less than the training procedure. Therefore, training the model costs $O(I \cdot S \cdot k \cdot d)$, where I is the number of iterations. In practical, I is always small for different datasets, in the range of $[5, 30]$.

3.4.3 Model Discussion

The ATTF model can be treated as a linear combination of two matrix factorization models which learn user preference and temporal effect respectively, as shown in Eq. (3.1). The first term depicts the user-POI interaction, which is similar as the low rank matrix factorization for the user-POI matrix through collaborative filtering

---

[2]http://www.mymedialite.net/index.html.

Algorithm 3: ATTF model learning algorithm

Input: Training tuples $\{(u_i, t_i, l_i)\}_{i=1,...,N}$

Output: $U^{(L)}, T_1^{(L)}, T_2^{(L)}, T_3^{(L)}, L^{(U)}, L^{(T)}$

Initialize $U^{(L)}, T_1^{(L)}, T_2^{(L)}, T_3^{(L)}, L^{(U)}, L^{(T)}$

Uniformly sample $\lfloor 100 * \sqrt{|\mathcal{U}|} \rfloor$ check-in tuples from D_S to generate D_e for loss calculation

for *iterations* **do**

    //S is the number of sampled check-ins

    for $i \in [1, S]$ **do**

        Draw (u, t, l_p) uniformly from training tuples

        // k is the number of negative samples

        for $n = 1, 2, \ldots, k$ **do**

            Draw l_n uniformly to form (u, t, l_p, l_n)

            $y_{u,t,l_p,l_n} \leftarrow y_{u,t,l_p} - y_{u,t,l_n}$

            $\delta \leftarrow 1 - \sigma(y_{u,t,l_p,l_n})$

            Update parameters according to Eq. (5.8)

        end

    end

    Estimate the loss defined on D_e

end

Return $U^{(L)}, T_1^{(L)}, T_2^{(L)}, T_3^{(L)}, L^{(U)}, L^{(T)}$

---

technique. The second term depicts the time-POI interaction, which acts like leveraging a latent factor model to describe the relations between time labels and POIs. Further, the aggregate operator $A(\cdot)$ combines several temporal factors together.

Two points are important to note for our model: (1) The ATTF model and the time labeling scheme are a general framework to subsume several temporal characteristics together. We take three common ones in this work, but it is easy to add others, e.g., different days in a month, workdays and vocations in a year. (2) Even though the model equation for ATTF in POI recommendation suffices to be expressed by a combination of two MF models, it is different from a simple ensemble of two MF model recommendation results because in our case the model parameters are learned jointly. Thus the learned parameters jointly represent the user preference and temporal effect. It better reflects the fact that user check-in behavior is a complex decision under many conditions.

The ATTF model can also be interpreted from the embedding neural network perspective. The embedding network, e.g., word2vec framework [20], has turned out to be a successful semi-supervised learning method in natural language processing [10, 22], network embedding [7, 23, 28], and recommendation systems [27, 29]. Moreover, recent studies [11, 12] show that the neural word embeddings can be treated as a kind of matrix factorization method [9]. This equivalence between neural embeddings and the latent factor models inspires us to understand our ATTF model from the embedding neural network perspective. Figure 3.5 demonstrates the equivalent embedding neural network for the ATTF model. The input layer is the one-hot representation for user, POI and temporal information. The second layer is the embedding layer, which projects the one-hot vector as a continuous latent

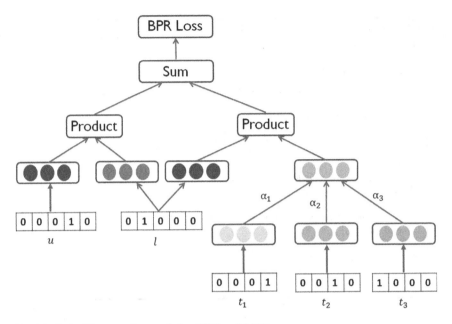

Fig. 3.5 Embedding neural network for ATTF model [36]

vector in the Euclidean subspace. Next, we exploit the product and sum operation to represent the check-in preference as $\langle U_u^{(L)}, L_l^{(U)} \rangle + \langle A(T_{1,t_1}^{(L)}, T_{2,t_2}^{(L)}, T_{3,t_3}^{(L)}), L_l^{(T)} \rangle$, equivalent to Eq. (3.1). Finally, we construct a BPR loss layer to learn the embedding representations.

3.5 Experiment

We conduct systematical experiments to seek the answers of the following questions: (1) how the proposed ATTF model performs comparing with state-of-the-art models? (2) whether the ATTF model is better than single temporal factor models? (3) how the parameters affect the model performance?

3.5.1 Data Description and Experimental Setting

Two real-world datasets are used in the experiment: one is Foursquare data from January 1, 2011 to July 31, 2011 provided in [6] and the other is Gowalla data from January 1, 2011 to September 31, 2011 in [35]. We filter the POIs checked-in by less than 5 users and then choose users who check-in more than 10 times as our samples.

Table 3.1 Statistics of datasets

Source	Foursquare	Gowalla
#users	10,180	3,318
#POIs	16,561	33,665
#check-ins	867,107	635,600
Avg. #check-ins each user	85.2	191.6
Avg. #POIs each user	24.3	104.1
Avg. #users each POI	14.9	10.3
Density[a]	0.0015	0.003

[a] Density means the fraction of checked-in entries over all entries in user-POI matrix

After the preprocessing, the datasets contain the statistical properties as shown in Table 3.1. We randomly choose 80% of each user's check-ins as training data, and the remaining 20% for test data. Moreover, we use each check-in (u, t, l) in training data to learn the latent features of user, time, and POI. Then given the (u, t), we estimate the score value of different candidate POIs, select the top N candidates, and compare them with check-in tuples in test data.

3.5.2 Performance Metrics

In this work, we leverage three metrics to evaluate the model performance–*precision*, *recall*, and *F-score*. The precision and recall in the top-K recommendation system are denoted as $P@K$ and $R@K$ respectively. $P@K$ measures the ratio of recovered POIs to the K recommended POIs, and $R@K$ means the ratio of recovered POIs to the set of POIs in the testing data. For each user $u \in \mathcal{U}$, $\mathcal{L}^T(u)$ denotes the set of correspondingly visited POIs in the test data, and $\mathcal{L}^R(u)$ denotes the set of recommended POIs. Then the definitions of $P@K$ and $R@K$ are formulated as follows

$$P@K = \frac{1}{|\mathcal{U}|} \sum_{u \in \mathcal{U}} \frac{|\mathcal{L}^R(u) \cap \mathcal{L}^T(u)|}{K}, \tag{3.9}$$

$$R@K = \frac{1}{|\mathcal{U}|} \sum_{u \in \mathcal{U}} \frac{|\mathcal{L}^R(u) \cap \mathcal{L}^T(u)|}{|\mathcal{L}^T(u)|}. \tag{3.10}$$

Further, *F-score* is the harmonic mean of precision and recall. So the *F-score* is defined as

$$F - score@K = \frac{2 * P@K * R@K}{P@K + R@K}. \tag{3.11}$$

3.5.3 Baselines

We compare our ATTF model with state-of-the-art collaborative filtering (CF) methods and POI recommendation methods incorporating temporal effect. Prior work [37, 38] indicates that treating the check-ins as implicit feedback is better to recommend POIs. Hence we exploit Weighted Regularized Matrix Factorization (WRMF) [8, 21] and Bayesian Personalized Ranking Matrix Factorization (BPR-MF) [24] as comparative CF methods. To illustrate the efficacy of our ATTF model, we compare it with LRT [4] and FPMC-LR [2] which are state-of-the-art POI recommendation methods incorporating temporal effect.

- **WRMF**. The WRMF model is designed for processing large scale implicit feedback data. We define the weight mapping of user u_i at POI l_j as $w_{i,j} = (1 + 10 \cdot C_{i,j})^{0.5}$, where $C_{i,j}$ is the check-in counts, following the setting in [18].
- **BPR-MF**. The BPR-MF model is a popular MF-based recommendation method to learn the pairwise relation, in which users prefer the observed items than the unobserved.
- **LRT**. The LRT model is designed to modeling the "non-uniformnes" and "consecutiveness" in a matrix factorization model for POI recommendation.
- **FPMC-LR**. The FPMC-LR model adds the Local Region constraint (i.e., geographical information) in the Factorized Personalized Markov Chain (FPMC) model [25]. FPMC-LR incorporates the geographical information and temporal consecutiveness through a local region constraint and the FPMC model respectively.

Moreover, to demonstrate the advantage of ATTF in aggregating several temporal latent factors, we also compare with three single temporal latent factor models: **TTFM**, **TTFW**, and **TTFH**. They are typically PITF model, that correspondingly considering the month, day of week, and hour as a temporal latent factor. Because these three models are the subset of our ATTF model, we attain their results by setting the corresponding weight as 1, and others as 0 in ATTF.

3.5.4 Experimental Results

Performance Comparison. In the following, we demonstrate the performance comparison on precision, recall and F-score. We set the latent factor dimension as 60 for all compared models. We leverage grid search method to find the best weights in ATTF model. α_1, α_2, and α_3 are constrained in the range of [0, 1]. In the grid search method, we first change α_1 from zero to one with step size 0.1. Then, for each α_1 value, for instance $\alpha_1 = 0.1$, we change α_2 from zero to $1 - \alpha_1$ with step size 0.1. α_3 can be calculated by $1 - \alpha_1 - \alpha_2$. The grid search method tries all value combinations with step size 0.1 satisfying the constraints $\alpha_1 + \alpha_2 + \alpha_3 = 1$, and $\alpha_1, \alpha_2, \alpha_3 >= 0$. As a result, the ATTF model on Foursquare data achieves the best

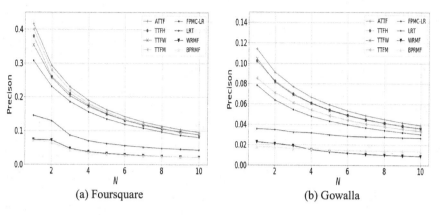

Fig. 3.6 Precision on Foursquare and Gowalla [36]

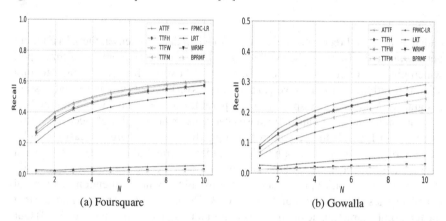

Fig. 3.7 Recall on Foursquare and Gowalla [36]

result when $\alpha_1 = 0.7$, $\alpha_2 = 0.1$, and $\alpha_3 = 0.2$, while the ATTF model on Gowalla data achieves the best when $\alpha_1 = 0.2$, $\alpha_2 = 0.1$, and $\alpha_3 = 0.7$.

Figures 3.6, 3.7 and 3.8 show the experimental results for Foursquare and Gowalla data on measurement precision, recall, and F-score respectively. We see that (1) Our proposed ATTF model outperforms state-of-the-art CF methods and POI recommendation models. Compared with the best state-of-the-art competitor in POI recommendation area (e.g., FPMC-LR), the ATTF model gains more than 20% enhancement on Foursquare data, and more than 36% enhancement on Gowalla data for all three measures, Precision@5, Recall@5, and F-score@5. We observe that models perform better on Foursquare data than Gowalla data, even though it is sparser. The reason lies in that Gowalla data contain much more POIs and a large candidate POI set makes the recommendation harder. (2) The ATTF model outperforms single temporal factor models. Compared with best single temporal factor model, the ATTF model gains about 3% enhancement on Foursquare data, and about 10% improvement on Gowalla

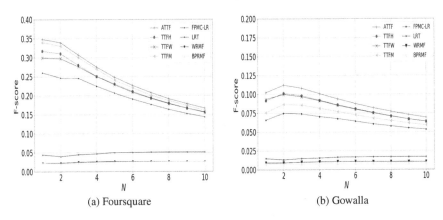

Fig. 3.8 F-score Foursquare and Gowalla [36]

data in a top-5 POI recommender system. So when data are denser, the ATTF model gets advantages. Because the ATTF model uses a tuple to represent a time spot, which gives more precise information. Dense data strengthen this precise labeling scheme. In addition, different weight assignments on both data give us two interesting insights: (i) When data are sparse, the temporal feature on month dominates the POI recommendation performance. Because check-ins on hour or day of week are sparse as shown in Fig. 3.3, then the corresponding characteristics are not easily caught. The Foursquare dataset has high weight on month temporal factor. However, when data are denser, check-ins on hour are not so sparse. So the temporal characteristic on hour of day becomes prominent. (ii) We usually pay much attention to temporal characteristics on hour of day and day of week. Our experimental results indicate that the temporal characteristic on month is important, especially for sparse data. (3) Our proposed ATTF model, single temporal factorization models (e.g., TTFM, TTFW, and TTFH), and FPMC-LR perform much better than other competitors, especially at recall measure. They try to recommend POIs at more specific situations, which is the key point to improve performance. Our models recommend a user POIs at some specific time, and FPMC-LR recommends POIs given a user's recent checked-in POIs; while, the other three models give general recommendations.

Parameter Effect. The regularization parameter and latent vector dimension are two important factors affecting the model performance. We explore how they affect the proposed model in the condition of other parameters fixed.

Figure 3.9 demonstrates the effect of regularization parameter on model performance. For simplicity, we set the same parameter λ for all latent vectors. The regularization part does not significantly affect the model. The model achieves the best performance at 0.001. With the increasing of λ, the performance decreases.

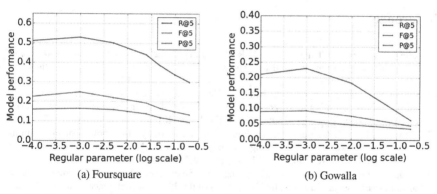

Fig. 3.9 The effect of regularization parameter λ [36]

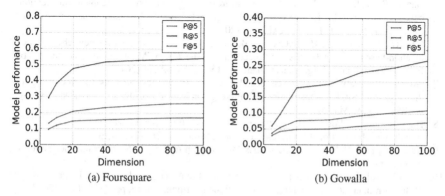

Fig. 3.10 The effect of latent factor dimension [36]

Figure 3.10 demonstrates how the latent vector dimension affects the model. The performance of ATTF steadily rises with the increase of latent vector dimension. For the trade-off of high performance and low computation cost, we suggest setting dimension $d = 60$.

3.6 Conclusion

In this chapter, we propose the ATTF model for POI recommendation. The proposed model introduces time factor to model the temporal effect in POI recommendation, subsuming all the three temporal properties: periodicity, consecutiveness, and non-uniformness. Moreover, the ATTF model captures the temporal influence at different time scales through aggregating several time factors' contributions. Experimental results on two real-world datasets show that the ATTF model outperforms state-of-the-art models. Our model is a general framework to aggregate several temporal characteristics at different scales.

References

1. Cheng, C., Yang, H., King, I., Lyu, M.R.: Fused matrix factorization with geographical and social influence in location-based social networks. In: Proceedings of the Twenty-Sixth AAAI Conference on Artificial Intelligence, pp. 17–23. AAAI Press (2012)
2. Cheng, C., Yang, H., Lyu, M.R., King, I.: Where you like to go next: successive point-of-interest recommendation. In: Proceedings of the Twenty-Third international joint conference on Artificial Intelligence, pp. 2605–2611. AAAI Press (2013)
3. Cho, E., Myers, S.A., Leskovec, J.: Friendship and mobility: user movement in location-based social networks. In: Proceedings of the 17th ACM SIGKDD International Conference on Knowledge Discovery and Data Mining, pp. 1082–1090. ACM (2011)
4. Gao, H., Tang, J., Hu, X., Liu, H.: Exploring temporal effects for location recommendation on location-based social networks. In: Proceedings of the 7th ACM conference on Recommender systems, pp. 93–100. ACM (2013)
5. Gao, H., Tang, J., Hu, X., Liu, H.: Content-aware point-of-interest recommendation on location-based social networks. In: Proceedings of the Twenty-Ninth AAAI Conference on Artificial Intelligence, pp. 1721–1727. AAAI Press (2015)
6. Gao, H., Tang, J., Liu, H.: gSCorr: modeling geo-social correlations for new check-ins on location-based social networks. In: Proceedings of the 21st ACM international conference on Information and knowledge management, pp. 1582–1586. ACM (2012)
7. Grover, A., Leskovec, J.: node2vec: Scalable feature learning for networks. In: Proceedings of the 22nd ACM SIGKDD International Conference on Knowledge Discovery and Data Mining, pp. 855–864. ACM (2016)
8. Hu, Y., Koren, Y., Volinsky, C.: Collaborative filtering for implicit feedback datasets. In: 2008 Eighth IEEE International Conference on Data Mining, pp. 263–272. IEEE (2008)
9. Koren, Y., Bell, R., Volinsky, C.: Matrix factorization techniques for recommender systems. Computer **42**(8), 30–37 (2009)
10. Le, Q.V., Mikolov, T.: Distributed representations of sentences and documents. In: Proceedings of the 31th International Conference on Machine Learning, vol. 14, pp. 1188–1196 (2014)
11. Levy, O., Goldberg, Y.: Neural word embedding as implicit matrix factorization. In: Advances in Neural Information Processing Systems, pp. 2177–2185 (2014)
12. Li, Y., Xu, L., Tian, F., Jiang, L., Zhong, X., Chen, E.: Word embedding revisited: A new representation learning and explicit matrix factorization perspective. In: Proceedings of the 25th International Joint Conference on Artificial Intelligence, pp. 3650–3656. AAAI Press (2015)
13. Lian, D., Ge, Y., Zhang, F., Yuan, N.J., Xie, X., Zhou, T., Rui, Y.: Content-aware collaborative filtering for location recommendation based on human mobility data. In: 2015 IEEE International Conference on Data Mining (ICDM), pp. 261–270. IEEE (2015)
14. Lian, D., Zhao, C., Xie, X., Sun, G., Chen, E., Rui, Y.: GeoMF: Joint geographical modeling and matrix factorization for point-of-interest recommendation. In: ACM SIGKDD International Conference on Knowledge Discovery and Data Mining, pp. 831–840. ACM (2014)
15. Liu, P., Qiu, X., Huang, X.: Learning context-sensitive word embeddings with neural tensor skip-gram model. In: Proceedings of the 25th international Joint Conference on Artificial Intelligence. AAAI Press (2015)
16. Liu, X., Liu, Y., Aberer, K., Miao, C.: Personalized point-of-interest recommendation by mining users' preference transition. In: Proceedings of the 22nd ACM international conference on Information & Knowledge Management, pp. 733–738. ACM (2013)
17. Liu, Y., Liu, Z., Chua, T.S., Sun, M.: Topical word embeddings. In: Proceedings of the 29th AAAI Conference on Artificial Intelligence. AAAI Press (2015)
18. Liu, Y., Wei, W., Sun, A., Miao, C.: Exploiting Geographical Neighborhood Characteristics for Location Recommendation. In: ACM International Conference on Conference on Information and Knowledge Management, pp. 739–748. ACM (2014)

19. Ma, H., Zhou, D., Liu, C., Lyu, M.R., King, I.: Recommender systems with social regularization. In: Proceedings of the fourth ACM International Conference on Web Search and Data Mining, pp. 287–296. ACM (2011)

20. Mikolov, T., Sutskever, I., Chen, K., Corrado, G.S., Dean, J.: Distributed representations of words and phrases and their compositionality. In: Advances in Neural Information Processing Systems, pp. 3111–3119 (2013)

21. Pan, R., Zhou, Y., Cao, B., Liu, N.N., Lukose, R., Scholz, M., Yang, Q.: One-class collaborative filtering. In: Proceedings of the 2008 Eighth IEEE International Conference on Data Mining, pp. 502–511. IEEE Computer Society (2008)

22. Pennington, J., Socher, R., Manning, C.: Glove: Global vectors for word representation. In: Proceedings of the 2014 Conference on Empirical Methods in Natural Language Processing (EMNLP), vol. 14, pp. 1532–1543 (2014)

23. Perozzi, B., Al-Rfou, R., Skiena, S.: Deepwalk: Online learning of social representations. In: Proceedings of the 20th ACM SIGKDD International Conference on Knowledge Discovery and Data Mining, pp. 701–710. ACM (2014)

24. Rendle, S., Freudenthaler, C., Gantner, Z., Schmidt-Thieme, L.: BPR: Bayesian personalized ranking from implicit feedback. In: Proceedings of the Twenty-Fifth Conference on Uncertainty in Artificial Intelligence, pp. 452–461. AUAI Press (2009)

25. Rendle, S., Freudenthaler, C., Schmidt-Thieme, L.: Factorizing Personalized Markov Chains for Next-basket Recommendation. In: Proceedings of the 19th International Conference on World Wide Web, pp. 811–820. ACM (2010)

26. Rendle, S., Schmidt-Thieme, L.: Pairwise interaction tensor factorization for personalized tag recommendation. In: Proceedings of the third ACM International Conference on Web Search and Data Mining, pp. 81–90. ACM (2010)

27. Tang, D., Qin, B., Liu, T., Yang, Y.: User modeling with neural network for review rating prediction. In: Proceedings of the 24th International Conference on Artificial Intelligence, pp. 1340–1346. AAAI Press (2015)

28. Tang, J., Qu, M., Wang, M., Zhang, M., Yan, J., Mei, Q.: Line: Large-scale information network embedding. In: Proceedings of the 24th International Conference on World Wide Web, pp. 1067–1077. ACM (2015)

29. Xie, M., Yin, H., Wang, H., Xu, F., Chen, W., Wang, S.: Learning Graph-based POI Embedding for Location-based Recommendation. In: Proceedings of the 25th ACM International on Conference on Information and Knowledge Management, pp. 15–24. ACM (2016)

30. Ye, M., Yin, P., Lee, W.C., Lee, D.L.: Exploiting geographical influence for collaborative point-of-interest recommendation. In: Proceedings of the 34th International ACM SIGIR Conference on Research and Development in Information Retrieval, pp. 325–334. ACM (2011)

31. Yin, H., Sun, Y., Cui, B., Hu, Z., Chen, L.: Lcars: A location-content-aware recommender system. In: ACM SIGKDD International Conference on Knowledge Discovery and Data Mining, pp. 221–229. ACM (2013)

32. Yuan, Q., Cong, G., Ma, Z., Sun, A., Thalmann, N.M.: Time-aware point-of-interest recommendation. In: Proceedings of the 36th International ACM SIGIR Conference on Research and Development in Information Retrieval, pp. 363–372. ACM (2013)

33. Zhang, J.D., Chow, C.Y.: iGSLR: personalized geo-social location recommendation: a kernel density estimation approach. In: Proceedings of the 21st ACM SIGSPATIAL International Conference on Advances in Geographic Information Systems, pp. 334–343. ACM (2013) N

34. Zhang, J.D., Chow, C.Y.: GeoSoCa: Exploiting geographical, social and categorical correlations for point-of-interest recommendations. In: Proceedings of the 38th International ACM SIGIR Conference on Research and Development in Information Retrieval, pp. 443–452. ACM (2015)

35. Zhao, S., King, I., Lyu, M.R.: Capturing geographical influence in poi recommendations. In: International Conference on Neural Information Processing, pp. 530–537. Springer (2013)

36. Zhao, S., King, I., Lyu, M.R.: Aggregated Temporal Tensor Factorization Model for Point-of-Interest Recommendation. Neural Process. Lett. (2017). https://doi.org/10.1007/s11063-017-9681-8

37. Zhao, S., Zhao, T., King, I., Lyu, M.R.: Geo-Teaser: Geo-Temporal Sequential Embedding Rank for Point-of-interest Recommendation. In: Proceedings of the 26th International Conference on World Wide Web Companion, pp. 153–162. International World Wide Web Conferences Steering Committee (2017)
38. Zhao, S., Zhao, T., Yang, H., Lyu, M.R., King, I.: STELLAR: Spatial-Temporal Latent Ranking for Successive Point-of-Interest Recommendation. In: Thirtieth AAAI Conference on Artificial Intelligence, pp. 315–322 (2016)

Chapter 4
Geo-Teaser: Geo-Temporal Sequential Embedding Rank for POI Recommendation

Abstract This chapter proposes a **Geo-Te**mporal sequential **e**mbedding **r**ank (Geo-Teaser) model for POI recommendation. Inspired by the success of the word2vec framework to model the sequential contexts, a *temporal POI embedding* model is proposed to learn POI representations under some particular temporal state. The temporal POI embedding model captures the contextual check-in information in sequences and the various temporal characteristics on different days as well. Furthermore, a new way of incorporating the geographical influence into the pairwise preference ranking method through discriminating the unvisited POIs according to geographical information, is employed to develop a geographically hierarchical pairwise preference ranking model. Finally, a unified framework is proposed to recommend POIs combining these two models. Experimental results on two real-life datasets show that the Geo-Teaser model outperforms state-of-the-art models.

Keywords Poi recommendation · Geographical influence · Temporal influence Sequential modeling · Embedding learning

4.1 Introduction

LBSNs such as Foursquare and Facebook Places have become popular services to attract users sharing their check-in behaviors, making friends, and writing comments on POIs. With the prosperity of LBSNs, POI recommendation comes out to improve the user experience, which mines users' check-in sequences and recommends places where an individual has not been. POI recommendation not only helps users explore new interesting places in a city, but also facilitates business owners to launch advertisements to target customers. Due to the significance for users and businesses, POI recommendation has attracted much academic attention, and thus a bunch of methods has been proposed to enhance the POI recommendation system [1, 8, 31, 32].

Modeling the sequential pattern of user check-ins is necessary for POI recommendation. Because successive check-ins are usually correlated [2, 16, 30, 35]. Markov chain model, recurrent neural network, and the word2vec framework are used to model the check-in sequences in previous work. Studies in [16, 30, 35] exploit the

© The Author(s), under exclusive license to Springer Nature Singapore Pte Ltd., part of Springer Nature 2018
S. Zhao et al., *Point-of-Interest Recommendation in Location-Based Social Networks*, SpringerBriefs in Computer Science, https://doi.org/10.1007/978-981-13-1349-3_4

Markov chain model to capture the successive check-ins' transitive pattern. Besides, researchers in [2, 4, 40] use the latent factor model based on the Markov chain property to model the successive check-ins' correlations. Recently, inspired by the success of deep learning, the neural network has been used to model the check-in sequences. Liu et al. [15] employ the recurrent neural network (RNN) to find the sequential correlations. The work in [17] models the check-in sequences through the word2vec framework to capture the sequential contexts. Moreover, we observe that check-in sequences on different days naturally exhibit the various temporal characteristics. For example, users always check-in at POIs around offices on weekday while visit shopping malls on weekend. However, all previous sequential models ignore the various temporal characteristics, which motivates our model.

Inspired by the success of the word2vec framework to model the sequential contexts [17], we propose a temporal POI embedding model to capture the contextual check-in information and the various temporal characteristics as well. In [17], all POIs are built as the "corpus", each POI is treated as a "word", and a user's all sequential check-ins are treated as a "sentence". Then, the word2vec framework [21] can be used to learn the POI embeddings, which contain the contextual relationships of consecutively visited POIs, showing better performance than Markov chain model. Nevertheless, the learned POI embeddings for capturing the sequential contexts cannot subsume the various temporal characteristics on different days. Moreover, the geographical influence is not considered in [17]. Studies on user mobility data show that the geographical influence is the most significant factor for POI recommendation [31, 34, 37]. Therefore, the geographical influence is expected to be incorporated to improve the POI recommendation.

To sum up, we propose the Geo-Teaser model for POI recommendation, as shown in Fig. 4.1. On the one hand, we propose a temporal POI embedding model to capture the contextual check-in information and the various temporal characteristics as well. In particular, we treat one user's check-in sequence in one day as a "sentence". Then we consider each sequence under a specific temporal state and define the *temporal POI*, referring to a POI taking a specific temporal state as context. Further, we propose the temporal POI embedding model to learn POI representations and temporal state representations. On the other hand, we incorporate the geographical influence into a pairwise preference ranking model and develop a geographically hierarchical pairwise preference ranking model. Traditionally, we assume users prefer the visited POIs than the unvisited and establish a pairwise ranking model to learn user preference on POIs [12, 40]. Previous studies [1, 31] indicate that users prefer POIs that are geographically adjacent to their visited POIs. This geographical characteristic inspires us to boost the traditional pairwise ranking model through hierarchical pairwise preference relations that discriminate the unvisited POIs according to POIs' geographical information. Finally, we propose the Geo-Teaser model as a unified framework to recommend POIs combining the temporal POI embedding model and the geographically hierarchical pairwise ranking model.

To sum up, this chapter bases on the published work [39] with the following contributions.

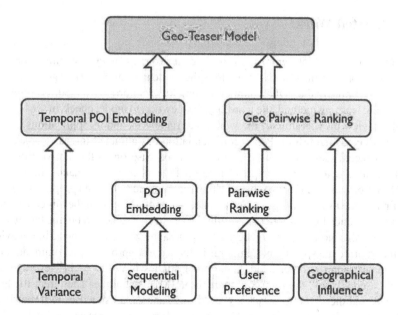

Fig. 4.1 Framework of the Geo-Teaser model

1. We propose the temporal POI embedding model, which captures the check-ins'
   sequential contexts and the various temporal characteristics on different days.
   In particular, we introduce the word2vec framework to project every POI as
   one object in an embedding space for learning the sequential relations among
   POIs. Furthermore, we learn the temporal POI representations from the check-in
   sequence under some specific temporal state.
2. We propose a new way to incorporate the geographical influence into the pairwise
   preference ranking method through discriminating the unvisited POIs according
   to geographical information. In particular, we define a hierarchical pairwise preference relation for each user check-in: the user prefers the visited POI than the
   unvisited neighboring POI, and the user prefers the unvisited neighboring POI
   than the unvisited non-neighboring POI. Then we learn the hierarchical pairwise
   preference to capture the geographical influence and user preference.
3. We propose the Geo-Teaser model as a unified framework combining the temporal POI embedding model and the geographically hierarchical pairwise preference ranking model. Experimental results on two real-life datasets show that
   the Geo-Teaser model outperforms state-of-the-art models. Compared with the
   best baseline competitor, the Geo-Teaser model improves at least 20% on both
   datasets for all metrics.

4.2 Related Work

In this section, we first demonstrate the recent progress of POI recommendation. Then we report how the prior work exploits the sequential influence and geographical influence to improve the POI recommendation. Since our proposed method adopts an embedding learning method, the word2vec framework, to model check-in sequences, we also review the literature of the word2vec framework and its applications.

POI Recommendation. POI recommendation has attracted intensive academic attention recently. Most of the proposed methods base on Collaborative Filtering (CF) techniques to learn user preference on POIs. On the one hand, the studies in [31, 33] employ the memory-based CF to recommend POIs. The proposed system first finds some users sharing the similar check-in preference with the target user and then recommends POIs where the "similar" users have checked-in but the target user has not. Furthermore, the researchers attempt to analyze the user check-in behavior and incorporate the spatial and temporal influence to improve the recommendation performance. On the other hand, some other studies in [1, 5, 6, 11] leverage the model-based CF, i.e., the Matrix Factorization (MF) technique. They treat the POI as "item" and the check-in frequency as "rating" and establish a user-POI matrix to recommend POIs using traditional MF models. Moreover, the researchers in [13, 19] observe that it is better to treat the check-ins as implicit feedback than explicit way, namely the check-ins are similar to clicks on Webs rather than the rating on Movies. They utilize the weighted regularized MF [9] to model this kind of implicit feedback. In addition, recent work in [12, 38, 40] employs pairwise ranking models to learn the user check-in as an implicit feedback and shows the advantages of ranking methods.

Sequential Modeling. Modeling the sequential pattern is important for POI recommendation. Most of the studies employ the Markov chain property in consecutive check-ins to capture the sequential pattern. We usually categorize the POI recommendation system as generic POI recommendation and successive POI recommendation by subtle differences in the recommendation task whether to be biased to the recent check-in. The successive POI recommendation is proposed to recommend POIs given the recent check-in, which naturally attempts to model the sequential pattern from successive check-ins [2, 4, 15, 34]. Also, researchers leverage the sequential modeling to improve the generic POI recommendation. The studies in [16, 30] learn the categories' transitive pattern in sequential check-ins. Zhang et al. [35] recommend POIs by learning the transitive probability through an additive Markov chain. Recently, inspired by the success of deep learning, the neural network has been used to model the check-in sequences. Liu et al. [15] employ the recurrent neural network (RNN) to find the sequential correlations among POIs. In the meantime, the work [17] models the check-in sequences through the word2vec framework [21] to capture the sequential contexts. The success in the prior work [15–17, 30, 35] motivates us to capture the sequential pattern in user check-ins to improve the generic POI recommendation. However, all previous sequential models ignore the various temporal characteristics. Hence, we propose a temporal POI embedding method to capture the sequential POIs' correlations under different temporal states.

Geographical Influence. Geographical influence plays an important role in POI recommendation. Compared with watching movies on Netflix and online shopping in Amazon, the check-in activity is limited to the physical constraint. Hence, the check-ins usually occur in the POIs nearby the user's home and working place. This observation motivates researchers to capture the geographical influence to improve the POI recommendation.

On the one hand, researchers attempt to establish geographical models to recommend POIs. First, researchers in [31, 33] discover that the distances for each pair of visited POIs in the LBSN follow the power law distribution after analyzing the geographical relations among visited POIs. Then, they propose a power law distribution model to fit the spatial relations among POIs and recommend POIs according to this kind of geographical influence [31, 33]. Moreover, researchers in [1, 3, 37] analyze each user's check-ins rather than all visited POIs and propose the Gaussian distribution based models to capture the geographical influence. Recently, Zhang et al. [34, 36] have observed that each user occupies a group of special parameters in the Gaussian mixture model. Then, they leverage the kernel density estimation to model each user's check-ins for personalization. On the other hand, instead of independently modeling the geographical influence, more researchers attempt to jointly model the geographical influence and other factors such as user preference and temporal influence together. The studies in [13, 19] incorporate the geographical influence into a weighted regularized MF model [9, 23] to learn the geographical influence and user preference together. Similar to [13, 19], we model the check-ins as a kind of implicit feedback. But we learn it through a Bayesian pairwise ranking method [26] due to its success in [40]. Furthermore, we propose a geographically hierarchical pairwise ranking model, which captures the geographical influence via discriminating the unvisited POIs according to their geographical information.

Embedding Learning. The word2vec framework [21] is an effective neural language model to learn embedding representations in word sequences. The key idea is to learn the sentence as the bag of words and represent the relations among words in the embedding subspace, such as "male" − "female" + "queen" = "king". The embedding learning technique in the word2vec framework attempts to capture the words' contextual correlations in sentences, showing better performance than the perspectives of word transitivity in sentences and word similarity. As a result, the embedding learning technique has been widely used in natural language processing recently [20, 22]. Afterwards, paragraph2vector [10] and other variants [14, 18] are proposed to enhance the word2vec framework for specific purposes. Since the efficacy of the framework in capturing the contextual correlations of items, the embedding technique based on the word2vec framework is employed to network embedding [24], as well as in user modeling [28] and item modeling [27]. To take the power of embedding learning for POI recommendation, Liu et al. [17] model the sequential contexts through a Skip-Gram model and achieves better performance than Markov chain model. Xie et al. [29] use similar embedding technique to recommend POIs. However, the previous work [17, 29] ignores two significant factors accounting for the check-in activity, the various temporal characteristics and geographical influence. To incorporate these two factors, we propose the Geo-Teaser model.

4.3 Data Description and Analysis

In this section, we first introduce two real-world LBSN datasets and then conduct the empirical analysis to explore the properties of check-in sequences in one day.

4.3.1 Data Description

We use two check-in datasets crawled from real-world LBSNs for data analysis. One is collected from Foursquare provided in [7] and the other is Gowalla data provided in [37]. We preprocess the data by filtering the POIs checked-in less than five users and users whose check-ins are less than ten times. Then we keep the remaining users' check-in records from January 1, 2011 to July 31, 2011. After the preprocessing, the datasets contain the statistical properties as shown in Table 4.1.

4.3.2 Empirical Analysis

We conduct data analysis to answer the following two questions: (1) how POIs in sequences of one day correlate each other? (2) how check-in sequences perform on different days?

We investigate the correlations of POIs in sequences of one day, as shown in Fig. 4.2. To calculate the correlation between two POIs, we construct the user-POI matrix according to the check-in records. Then, we measure the correlation of a

Table 4.1 Data statistics

	Foursquare	Gowalla
#users	10,034	3,240
#POIs	16,561	33,578
#check-ins	865,647	556,453
Avg. #check-ins of each user	86.3	171.7
Avg. #POIs for each user	24.6	95.4
Avg. #users for each POI	14.9	9.2
Density	0.0015	0.0028

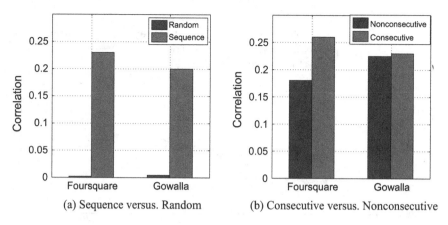

Fig. 4.2 POI correlation in sequences [39]

POI pair regarding the Jaccard similarity of those users who have checked-in at the two POIs. In Fig. 4.2a, we calculate the average correlation value of POI pairs in sequences for all users and compare it with the average correlation value of 5,000 random POI pairs. We observe that the correlation of POIs in sequences is much higher than random pairs by about 100 times for Foursquare and 50 times for Gowalla, which motivates the sequential modeling. In Fig. 4.2b, we compare the correlation of consecutive pairs with nonconsecutive pairs in sequences. Take a sequence of (l_1, l_2, l_3) as an example, (l_1, l_2) and (l_2, l_3) are consecutive pairs, and (l_1, l_3) is a nonconsecutive pair. We also calculate the average value of all sequences for all users to make the comparison. We observe that the nonconsecutive pairs contain comparable correlation to the consecutive pairs. Hence, not only consecutive POIs are highly correlated [2, 40], all POIs in a sequence are highly correlated with a contextual property. Accordingly, it is not satisfactory to only model the consecutive check-ins' transitive probability by Markov chain model or the consecutive check-ins' correlation by tensor factorization. This observation motivates us to model the whole sequence through the word2vec framework.

We explore how the various temporal characteristics on different days affect the user's check-in behavior. Previous work [38, 40] shows that user check-ins exhibit different patterns on different days, especially for working days and weekends. Figure 4.3 demonstrates the number of cumulated check-ins for all users at different hours on different days of a week, from Monday to Sunday. From the statistics of cumulated check-ins in Fig. 4.3, we observe the day of week check-in pattern at different hours: users take more check-ins in the late afternoon and the evening from 16:00 p.m. to 3:00 a.m. on weekends than the weekdays. Hence, Saturday and Sunday take the similar pattern, while the days from Monday to Friday take the similar pattern that is different from the weekends. We may infer that weekday and weekend exert two types of effects on the user's check-in behavior. Therefore, modeling the sequence pattern should contain this temporal feature.

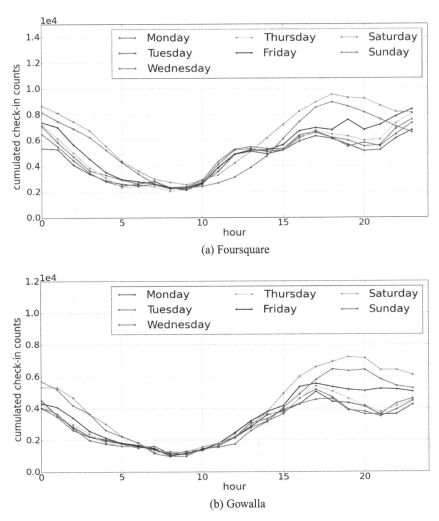

Fig. 4.3 Check-in pattern at different hours over day of week [39]

4.4 Method

In this section, we first propose the temporal POI embedding model to capture the various temporal characteristics for sequential modeling. Next, we demonstrate the geographically hierarchical pairwise preference ranking model. Then, we propose the Geo-Teaser model as a unified framework to recommend POIs combining the temporal POI embedding model and the geographically hierarchical pairwise preference ranking model. Finally, we show the learning procedures for the Geo-Teaser model.

4.4.1 Temporal POI Embedding

We propose a temporal POI embedding method to learn the sequential pattern, which captures POIs' contextual information from user check-in sequences and as well as the various temporal characteristics. Different from the work [17] that treats a user's all check-ins as a "sentence", we treat a user's check-ins of one day as a "sentence". Because consecutive check-ins on different days may span a long time and be not highly correlated. Further, we assume that check-in sequences on different days exhibit various temporal characteristics. Then, we learn POI embeddings in a sequence with some specific temporal state.

To better describe the model, we present some basic concepts as follows.

Definition 4.1 (*Check-in*) A check-in is a triple $\langle u, l, t \rangle$ that depicts a user u visiting POI l at time t.

Definition 4.2 (*Check-in sequence*) A check-in sequence is a set of check-ins of user u in one day, denoted as $S_u = \{\langle l_1, t_1 \rangle, \ldots, \langle l_n, t_n \rangle\}$, where t_1 to t_n belong to the same day. For simplicity, we denote $S_u = \{l_1, \ldots, l_n\}$.

Definition 4.3 (*Target POI and context POI*) In a sequence S_u, the chosen l_i is the target POI and other POIs in S_u are context POIs.

We propose the temporal POI embedding model based on the Skip-Gram model [21]. As shown in Fig. 4.4, we learn the representations of context POIs from l_{i-k} to l_{i+k} given a target POI l_i and the sequence temporal state t_s. Here k is a parameter to control the context window size. In addition, the temporal state t_s is composed of two options, weekday and weekend. Because we want to discriminate weekday and weekend, which depict the various temporal characteristics on day level as shown in Fig. 4.3. Formally, given a sequence S_u and its temporal state

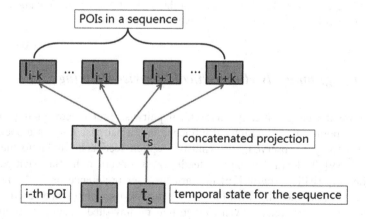

Fig. 4.4 Temporal POI embedding model [39]

t_s, our model attempts to learn the temporal POI embeddings through maximizing the following function,

$$\mathscr{L}_{TPE} = \sum_{S_u \in S} \frac{1}{|S_u|} \sum_{l_i \in S_u} \sum_{-k \leq c \leq k, c \neq 0} \left(\log \Pr(l_{i+c}|l_i, t_s) \right), \tag{4.1}$$

where S is a set containing all sequences S_u for all users. $\mathscr{L}_{TPE}$ aims to maximize the context POI's conditional occurrence likelihood for all sequences.

Furthermore, we formulate the probability $\Pr(l_{i+c}|l_i, t_s)$ using a softmax function. For better description, we introduce two symbols, defined as follows: $\hat{\mathbf{l}}_c' = \mathbf{l}_c' \oplus \mathbf{l}_c'$, $\mathbf{l}_i^t = \mathbf{l}_i \oplus \mathbf{t}_s$, where $\oplus$ is the concatenation operator, and $\mathbf{l}_c', \mathbf{l}_i$, and $\mathbf{t}_s$ are latent vectors of output layer context POI, target POI, and temporal state, respectively. Thus, we get $\hat{\mathbf{l}}_c' \cdot \mathbf{l}_i^t = \mathbf{l}_c' \cdot \mathbf{l}_i + \mathbf{l}_c' \cdot \mathbf{t}_s$. Therefore, the probability $\Pr(l_{i+c}|l_i, t_s)$ can be formulated as,

$$\Pr(l_{i+c}|l_i, t_s) = \frac{\exp(\hat{\mathbf{l}}_c' \cdot \mathbf{l}_i^t)}{\sum_{l_i \in L} \exp(\hat{\mathbf{l}}_c' \cdot \mathbf{l}_i^t)}. \tag{4.2}$$

As the size of set L in Eq. (4.2) is large, we exploit the negative sampling technique [21] to learn the model efficiently. Then, the objective function can be formulated in a new form easier to optimize,

$$\mathscr{L}_{TPE} = \sum_{S_u \in S} \frac{1}{|S_u|} \sum_{l_i \in S_u} \sum_{-k \leq c \leq k, c \neq 0} \left(\log \sigma(\hat{\mathbf{l}}_c' \cdot \mathbf{l}_i^t) + \sum_h E_{k'} \log \sigma(-\hat{\mathbf{l}}_{k'}' \cdot \mathbf{l}_i^t) \right), \tag{4.3}$$

where $l_{k'}$ is the sampled negative POI, h is the number of negative samples, $\sigma(\cdot)$ is the sigmoid function, and $E(\cdot)$ means to calculate the expectation value for all generated negative samples. Here we adopt the same strategy in [21], namely using a unigram distribution, to draw the negative samples.

4.4.2 Geographically Hierarchical Pairwise Ranking

We propose the geographically hierarchical pairwise preference ranking model, which incorporates the geographical influence into a pairwise ranking model. The check-in activity is observed as a kind of implicit feedback similar to the web clicks [13, 19]. To learn this implicit feedback, we leverage the Bayesian personalized ranking (BPR) criteria [26] to learn the user preference on POIs. BPR is a pairwise ranking model, which learns the pairwise user preference based on the assumption that users prefer the visited POIs than the unvisited. In our geographically hierarchical pairwise ranking model, we discriminate the unvisited POIs using POIs' geographical information. Previous studies [1, 33, 37] observe that users prefer the POIs nearby the visited than POIs far away, we can discriminate the unvisited POIs and define *neighboring POI* and *non-neighboring POI* as follows.

Definition 4.4 (*Neighboring POI and non-neighboring POI*) For each check-in $\langle u, l_i \rangle$, the neighboring POI is the POI whose distance from l_i is less than or equal to a threshold s, while the non-neighboring POI is the POI whose distance is more than s.

Furthermore, for each check-in $\langle u, l_i \rangle$, we define a hierarchical pairwise preference relation: the user prefers the visited POI l_i than the unvisited neighboring POI l_{ne}, and prefers the unvisited neighboring POI l_{ne} than the unvisited non-neighboring POI l_{nn}. Denote $d(l_i, l_j)$ as the distance of two POIs l_i and l_j, we represent the hierarchical pairwise preference relation for check-in $\langle u, l_i \rangle$ as follows,

$$l_i >_{u, d(l_i, l_{ne}) \leq s} l_{ne} \vee l_{ne} >_{u, d(l_i, l_{nn}) > s} l_{nn}. \tag{4.4}$$

Suppose L is the set of POIs, and L_u is the visited POIs of user u, the hierarchical pairwise preference relation set for a sequence S_u satisfying Eq. (4.4) is defined as follows,

$$D_{S_u} = \{(u, l_i, l_{ne}) \vee (u, l_{ne}, l_{nn}) | l_i \in S_u, d(l_i, l_{ne}) \leq s,$$
$$d(l_i, l_{nn}) > s, l_{ne}, l_{nn} \in L \setminus L_u\}. \tag{4.5}$$

Now learning the geographically hierarchical pairwise ranking model is equivalent to model the preference relations in D_{S_u}. Here we employ the MF model to formulate the preference score function. We use $\mathbf{l}_i^t = \mathbf{l}_i \oplus \mathbf{t}_s$ to represent the temporal POI latent vector, which is consistent with the temporal POI embedding model. In addition, we define $\hat{\mathbf{u}} = \mathbf{u} \oplus \mathbf{u}$, then the score function can be formulated as,

$$f(u, t_s, l_i) = \hat{\mathbf{u}} \cdot \mathbf{l}_i^t. \tag{4.6}$$

Next, we use the sigmoid function to formulate the pairwise preference probability. Suppose $\text{Pr}(l_i >_u l_n)$ denotes the probability of user u prefers POI l_i than l_n, and $\sigma(\cdot)$ is the sigmoid function. Then, each pair in the preference set can be formulated as,

$$\text{Pr}(l_i >_u l_n) = \sigma(f(u, t_s, l_i) - f(u, t_s, l_n)) = \sigma(\mathbf{u} \cdot (\mathbf{l}_i - \mathbf{l}_n)). \tag{4.7}$$

Thus, learning the geographically hierarchical pairwise ranking model is equivalent to maximize the following function,

$$\mathscr{L}_{GPR} = \sum_{S_u \in S} \sum_{(u, l_i, l_n) \in D_{S_u}} \log \sigma(\mathbf{u} \cdot (\mathbf{l}_i - \mathbf{l}_n)), \tag{4.8}$$

where S is a set containing all sequences S_u for all users and D_{S_u} is hierarchical pairwise preference relations on sequence S_u.

4.4.3 Geo-Teaser Model

We propose the Geo-Teaser model as a unified framework to recommend POIs combining the temporal embedding model and the pairwise ranking model. Learning the Geo-Teaser model is equivalent to maximize $\mathscr{L}_{TPE}$ and $\mathscr{L}_{GPR}$ together,

$$\mathscr{O} = \underset{\mathbf{U,L,T}}{\arg\max}\, \alpha \cdot \mathscr{L}_{TPE} + \beta \cdot \mathscr{L}_{GPR}, \tag{4.9}$$

where α and β are the hyperparameters to trade-off the sequential modeling and the preference learning modules. We expect to obtain the user, POI, and temporal state representations through learning the temporal POI embeddings and geographically pairwise preference relations in the Geo-Teaser model.

Substituting $\mathscr{L}_{TPE}$ and $\mathscr{L}_{GPR}$ with Eq. (4.3) and Eq. (4.8) respectively, then we can learn the Geo-Teaser model through the following objective function,

$$\begin{aligned}
\underset{\mathbf{U,L,T}}{\arg\max} \sum_{S_u \in S} \sum_{l_i \in S_u} \Big(\sum_{-k \leq c \leq k, c \neq 0} &\alpha \log \sigma(\mathbf{l}'_c \cdot \mathbf{l}_i) + \\
&\sum_h \alpha E_{k'} \log \sigma(-\mathbf{l}'_{k'} \cdot \mathbf{l}_i) + \\
&\sum_{D_{S_u}} \beta \log(\sigma(\mathbf{u} \cdot (\mathbf{l}_i - \mathbf{l}_n)))\Big).
\end{aligned} \tag{4.10}$$

4.4.4 Learning

We use an alternate iterative update procedure and employ stochastic gradient descent (SGD) to learn the objective function. To learn the model, for each sampled training instance, we separately calculate the derivatives for $\mathscr{L}_{TPE}$ and $\mathscr{L}_{GPR}$, and then update the corresponding parameters along the ascending gradient direction,

$$\Theta^{t+1} = \Theta^t + \eta \times \frac{\partial \mathscr{O}(\Theta)}{\partial \Theta}, \tag{4.11}$$

where Θ is the training parameter and η is the learning rate.

Specifically, for a check-in $\langle u, l_i \rangle$, we calculate the stochastic gradient decent for $\mathscr{L}_{TPE}$. First, we get the updating rule for the context POI l_c,

$$\begin{aligned}
\mathbf{l}_i &\leftarrow \mathbf{l}_i + \alpha \eta (1 - \sigma(\hat{\mathbf{l}}'_c \cdot \mathbf{l}'_i)) \mathbf{l}'_c \\
\mathbf{t}_i &\leftarrow \mathbf{t}_i + \alpha \eta (1 - \sigma(\hat{\mathbf{l}}'_c \cdot \mathbf{l}'_i)) \mathbf{l}'_c \\
\mathbf{l}'_c &\leftarrow \mathbf{l}'_c + \alpha \eta (1 - \sigma(\hat{\mathbf{l}}'_c \cdot \mathbf{l}'_i))(\mathbf{l}_i + \mathbf{t}_i).
\end{aligned} \tag{4.12}$$

Algorithm 4: Learning algorithm for the Geo-Teaser model

Input: S

Output: $\mathbf{U}, \mathbf{L}, \mathbf{T}$

1 Initialize $\mathbf{U}, \mathbf{L}, \mathbf{L}'$, and $\mathbf{T}$ (uniformly at random)

2 **for** *iterations* **do**

3   **for** $S_u \in S$ **do**

4   **for** $\langle u, l_i \rangle \in S_u$ **do**

5   **for** *each context POI* l_c **do**

6              $\mathbf{l}_i \leftarrow \mathbf{l}_i + \alpha\eta(1 - \sigma(\hat{\mathbf{l}}'_c \cdot \mathbf{l}^t_i))\mathbf{l}'_c$

7              $\mathbf{t}_i \leftarrow \mathbf{t}_i + \alpha\eta(1 - \sigma(\hat{\mathbf{l}}'_c \cdot \mathbf{l}^t_i))\mathbf{l}'_c$

8              $\mathbf{l}'_c \leftarrow \mathbf{l}'_c + \alpha\eta(1 - \sigma(\hat{\mathbf{l}}'_c \cdot \mathbf{l}^t_i))(\mathbf{l}_i + \mathbf{t}_i)$

9   **for** $\{k' \sim P_{nc_c}\}$ **do**

10                $\mathbf{l}_i \leftarrow \mathbf{l}_i - \alpha\eta\sigma(\hat{\mathbf{l}}'_{k'} \cdot \mathbf{l}^t_i)\mathbf{l}'_{k'}$

11                $\mathbf{t}_i \leftarrow \mathbf{t}_i - \alpha\eta\sigma(\hat{\mathbf{l}}'_{k'} \cdot \mathbf{l}^t_i)\mathbf{l}'_{k'}$

12                $\mathbf{l}'_{k'} \leftarrow \mathbf{l}'_{k'} - \alpha\eta\sigma(\hat{\mathbf{l}}'_{k'} \cdot \mathbf{l}^t_i)(\mathbf{l}_i + \mathbf{t}_i)$

13   **end**

14   **end**

15   Uniformly sample m unvisited POIs

16   **for** $(u, l_i, l_{ne}) \in D_m$ **do**

17          $\delta = 1 - \sigma(\mathbf{u} \cdot \mathbf{l}_i - \mathbf{u} \cdot \mathbf{l}_{ne})$

18          $\mathbf{u} \leftarrow \mathbf{u} + \beta\eta\delta(\mathbf{l}_i - \mathbf{l}_{ne})$

19          $\mathbf{l}_i \leftarrow \mathbf{l}_i + \beta\eta\delta\mathbf{u} \, ; \mathbf{l}_{ne} \leftarrow \mathbf{l}_{ne} - \beta\eta\delta\mathbf{u}$

20   **end**

21   **for** $(u, l_{ne}, l_{nn}) \in D_m$ **do**

22          $\delta = (1 - \sigma(\mathbf{u} \cdot \mathbf{l}_{ne} - \mathbf{u} \cdot \mathbf{l}_{nn}))$

23          $\mathbf{u} \leftarrow \mathbf{u} + \beta\eta\delta(\mathbf{l}_{ne} - \mathbf{l}_{nn})$

24          $\mathbf{l}_{ne} \leftarrow \mathbf{l}_{ne} + \beta\eta\delta\mathbf{u} \, ; \mathbf{l}_{nn} \leftarrow \mathbf{l}_{nn} - \beta\eta\delta\mathbf{u}$

25   **end**

26   **end**

27 **end**

28 **end**

Then, we update the negative sample l'_k as follows,

$$\mathbf{l}_i \leftarrow \mathbf{l}_i - \alpha\eta\sigma(\hat{\mathbf{l}}'_{k'} \cdot \mathbf{l}^t_i)\mathbf{l}'_{k'}$$
$$\mathbf{t}_i \leftarrow \mathbf{t}_i - \alpha\eta\sigma(\hat{\mathbf{l}}'_{k'} \cdot \mathbf{l}^t_i)\mathbf{l}'_{k'} \tag{4.13}$$
$$\mathbf{l}'_{k'} \leftarrow \mathbf{l}'_{k'} - \alpha\eta\sigma(\hat{\mathbf{l}}'_{k'} \cdot \mathbf{l}^t_i)(\mathbf{l}_i + \mathbf{t}_i).$$

To update $\mathscr{L}_{GPR}$, we calculate the stochastic gradient decent for each preference pair (u, l_i, l_n) in D_{S_u}.[1] Denote $\delta = 1 - \sigma(\mathbf{u} \cdot \mathbf{l}_i - \mathbf{u} \cdot \mathbf{l}_n)$, we update the parameters as follows,

---

[1]The pair of (u, l_i, l_n) happens in two cases: (u, l_i, l_{ne}) and (u, l_{ne}, l_{nn}) as shown in Algorithm 4.

$$\mathbf{u} \leftarrow \mathbf{u} + \beta \eta \delta (\mathbf{l}_i - \mathbf{l}_n)$$
$$\mathbf{l}_i \leftarrow \mathbf{l}_i + \beta \eta \delta \mathbf{u} \tag{4.14}$$
$$\mathbf{l}_n \leftarrow \mathbf{l}_n - \beta \eta \delta \mathbf{u}.$$

Algorithm 4 shows the details of learning the Geo-Teaser model. S is the set of all sequences, and S_u is a sequence of user u. U, L, and T are feature matrices of the user, POI, and temporal state. $\mathbf{L}'$, an assistant learning parameter, is the output layer POI matrix in Skip-Gram model. We use the standard way [21] to learn the POI representations in the sequences, as shown from line 5 to line 14 in Algorithm 4. Next, we exploit the Bootstrap sampling to generate m unvisited POIs and then classify the unvisited POIs as neighboring POIs and non-neighboring POIs according to their distances from the visited POI l_i. Then, we establish the pairwise preference set D_m for each check-in $\langle u, l_i \rangle$. Here $D_m = \{(u, l_i, l_{ne}) \vee (u, l_{ne}, l_{nn}) | d(l_i, l_{ne}) \leq s, d(l_i, l_{nn}) > s, l_{ne}, l_{nn} \in L \setminus L_u\}$. Then we learn the parameters for each instance in D_m, shown from line 15 to line 25 in Algorithm 4.

After learning the Geo-Teaser model, we get the latent feature representations of users, POIs, and temporal states. Then, we can estimate the check-in possibility of user u over a candidate POI l at temporal state t_s according to the preference score function. Furthermore, we use the Eq. (4.6) for score estimation. Finally, we rank the candidate POIs and select the top N POIs with the highest estimated possibility values for each user.

Scalability. For one check-in, learning the temporal embedding model costs $O(k \cdot h \cdot d)$, where k, h, and d denote the context window size, the number of negative samples, and the latent vector dimension, respectively. For the pairwise preference learning from line 15 to 25 in Algorithm 4, we sample m unvisited POIs, which can generate maximum $O(m^2)$ pairwise preference tuples. For each check-in, the learning procedures cost $O(m^2 \cdot d)$. Therefore, the complexity of our model is $O((k \cdot h + m^2) \cdot d \cdot |C|)$, where C is the set of all check-ins. For k, h, m, and d are fixed hyperparameters, the proposed model can be treated as linear in $O(|C|)$. Furthermore, in order to make our model more efficient, we turn to the asynchronous stochastic gradient descent (ASGD) [25] and parallelly run the algorithm in an unlock way. As the check-in frequency distribution of POIs in LBSNs follows a power law [31], this results in a long tail of infrequent POIs, which guarantees to employ the ASGD to parallel the parameter updates.

4.5   Experimental Evaluation

We conduct experiments to seek the answers to the following questions: (1) how the Geo-Teaser model performs comparing with state-of-the-art recommendation methods? (2) how each component (i.e., the various temporal characteristics and geographical influence) affects the model performance? (3) how the parameters affect the model performance?

4.5.1 Experimental Setting

Two real-world datasets are used in the experiment: one is from Foursquare provided in [7] and the other is from Gowalla in [37]. Table 4.1 demonstrates the statistical information of the datasets. In order to make our model satisfactory to the scenario of recommending for future check-ins, we choose the first 80% of each user's check-ins as training data, the remaining 20% for test data, following [2, 35].

4.5.2 Performance Metrics

In this work, we compare the model performance through *precision* and *recall*, which are generally used to evaluate a POI recommendation system [5, 12]. To evaluate a top-N recommendation system, we denote the precision and recall as P@N and R@N, respectively. In our POI recommendation task, P@N measures the ratio of recovered POIs to the N recommended POIs, and R@N means the ratio of recovered POIs to the set of POIs in the test data. Then we calculate the average precision and recall over all users for evaluation. Supposing $L_{visited}$ denotes the set of correspondingly visited POIs in the test data, and $L_{N,rec}$ denotes the set of recommended POIs, the definitions of P@N and R@N are formulated as follows,

$$P@N = \frac{1}{|U|} \sum_{u \in U} \frac{|L_{visited} \cap L_{N,rec}|}{N}, \tag{4.15}$$

$$R@N = \frac{1}{|U|} \sum_{u \in U} \frac{|L_{visited} \cap L_{N,rec}|}{|L_{visited}|}. \tag{4.16}$$

4.5.3 Model Comparison

Prior work [13, 19] observes that treating the check-ins as implicit feedback is better to model the user preference. Hence we compare our model with WRMF [9, 23] and BPRMF [26], which are state-of-the-art collaborative filtering models designed for capturing the implicit feedback. To illustrate the effectiveness of our model, we compare it with four state-of-the-art POI recommendation methods: LRT [5], LORE [35], Rank-GeoFM [12], and SG-CWARP [17].

- **BPRMF** [26]: **B**ayesian **P**ersonalized **R**anking **M**atrix **F**ac-torization (*BPRMF*) is a popular pairwise ranking method that models the implicit feedback data to recommend top-N items.
- **WRMF** [9, 23]: **W**eighted **R**egularized **M**atrix **F**actorization (*WRMF*) model is designed for implicit feedback ranking problem. We set the weight mapping

function of user u_i at POI l_j as $w_{i,j} = (1 + 10 \cdot C_{i,j})^{0.5}$, where $C_{i,j}$ is the number of check-ins, following the setting in [19].

- **LRT** [5]: **L**ocation **R**ecommendation framework with **T**emporal effects model (*LRT*) is a state-of-the-art POI recommendation method, which captures the temporal effect in POI recommendation.
- **LORE** [35]: *LORE* is state-of-the-art model that exploits the sequential influence for location recommendation. Compared with other work [2, 30], *LORE* employs the whole sequence's contribution, not only the successive check-ins sequential influence.
- **Rank-GeoFM** [12]: *Rank-GeoFM* is a ranking based geographical factorization method, which incorporates the geographical and temporal influence in a latent ranking model.
- **SG-CWARP** [17]. *SG-CWARP* is the latest work, which leverages the word2vec framework to model the check-ins for sequential contexts.

4.5.4   Experimental Results

In the following, we demonstrate the experimental results on precision and recall, denoted as P@N and R@N, for the top N POI recommendation task. Since the model comparison results are consistent with different values of N, e.g., 1, 5, 10, and 20, we show representative results at 5 and 10 following [5, 6]. All models achieve the best performances at appropriate parameter settings.

4.5.4.1   Performance Comparison

Figure 4.5 illustrates the experimental results of different models. We discover that the proposed Geo-Teaser model achieves better performance than all the baselines. Compared with Rank-GeoFM that is a state-of-the-art model incorporating the geographical influence and temporal influence, Geo-Teaser achieves improvements at least 28% on both datasets for all metrics. This verifies the effectiveness of our sequential modeling and as well as the validity of means for incorporating various temporal characteristics and geographical influence. SG-CWARP is the best baseline competitor, which verifies the advantage of modeling the sequential pattern through Skip-Gram model over Markov chain model, namely the LORE model. Our Geo-Teaser model outperforms the SG-CWARP at least 20% on both datasets for all metrics, which verifies our strategy of incorporating various temporal characteristics and geographical influence to improve POI recommendation. In addition, we observe that models perform better on Gowalla than Foursquare for *precision*, but worse for *recall*. The reason lies in that each user's test data size in Gowalla is bigger than Foursquare. As shown in Table 4.1, the average check-ins for each user in Gowalla

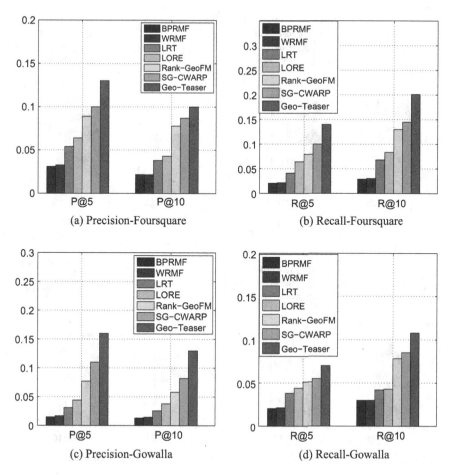

Fig. 4.5 Model comparison [39]

is about two times of Foursquare. According to the metrics in Eqs. (4.15) and (4.16), the result is reasonable.

4.5.4.2 Model Discussion

In this section, we explore how each component, i.e., the various temporal characteristics and geographical influence, affects the model performance. The Geo-Teaser model improves the SG-CWARP in two aspects, capturing the various temporal characteristics and geographical influence. Ignoring the various temporal characteristics and geographical influence, we propose the **SG-BPRMF** model as the basic version of our proposed Geo-Teaser model. The SG-BPRMF uses the Skip-Gram model to model the sequence and BPRMF to capture the user preference, which is equivalent to

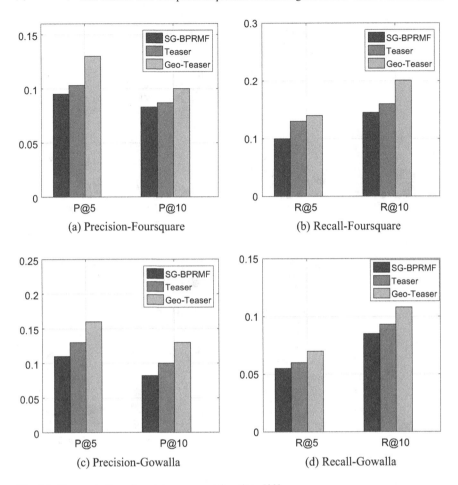

Fig. 4.6 Demonstration of model component functions [39]

SG-CWARP. Furthermore, we incorporate the various temporal characteristics into SG-BPRMF and propose the **Teaser** model. In the following, we compare the SG-BPRMF, Teaser, and Geo-Teaser to show how the various temporal characteristics and geographical influence affect the model.

Figure 4.6 shows the model performances. We observe that Teaser model improves SG-BPRMF at least about 10% on both datasets for all metrics, which indicates that incorporating the various temporal characteristics improves the model performance. Moreover, the Geo-Teaser model improves the Teaser model at least about 15% on both datasets. It means our strategy of incorporating geographical influence by discriminating the unvisited POIs is valid.

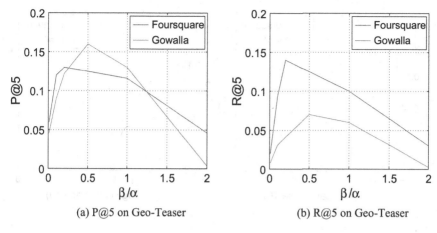

(a) P@5 on Geo-Teaser   (b) R@5 on Geo-Teaser

Fig. 4.7 Parameter effect on α and β [39]

4.5.4.3   Parameter Effect

In this section, we show how the three important hyperparameters, α, β, and s affect the model performance. α and β balance the sequential influence and the user preference. s shows the sensitivity of our geographical model.

We tune α and β to see how to trade-off the sequential modeling and user preference learning, shown in Fig. 4.7. Both α and β appear together with the learning rate η in the parameter update procedures. It is not necessary to separately tune the three parameters. We are able to absorb the learning rate η into α and β. In other words, we set $\alpha \leftarrow \alpha \cdot \eta$, $\beta \leftarrow \beta \cdot \eta$. We avoid to tune the learning rate η, but turn to control the update step size through tuning α and β. Hence α and β should be small enough to guarantee convergence. Assuming the same value for α and β, we tune α to change the learning rate. The model gets the best performance when $\alpha = 0.05$. Then we set $\alpha = 0.05$, and change β to see how the model performance varies with $\frac{\beta}{\alpha}$. Geo-Teaser attains the best performance if $\frac{\beta}{\alpha} \in [0.25, 0.5]$.

In the Geo-Teaser model, we classify the unvisited POIs as neighboring POIs and non-neighboring POIs to constitute a new preference set according to a threshold distance s. Here we choose different values of s to see how this parameter affects the model performance, as shown in Fig. 4.8. Here s is calculated in the kilometer. We observe that the Geo-Teaser model achieves the best performance at $s = 10$.

4.6   Conclusion

We propose the Geo-Teaser model for POI recommendation in this chapter. In particular, we propose the temporal POI embedding model to capture the check-ins' sequential contexts and the various temporal characteristics on different days. Moreover,

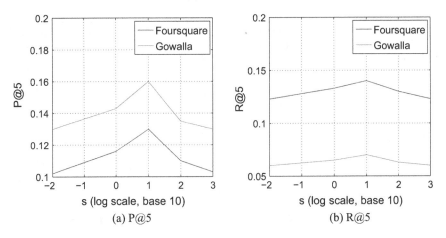

Fig. 4.8 Parameter effect on distance threshold s [39]

we propose the geographically hierarchical pairwise ranking model to improve the recommendation performance through incorporating geographical influence. Finally, we propose a unified framework combining the two parts to recommend POIs. Experimental results on two datasets, Foursquare and Gowalla, show that our model outperforms state-of-the-art models. The proposed Geo-Teaser model improves at least 20% on both datasets for all metrics compared with SG-CWARP model.

References

1. Cheng, C., Yang, H., King, I., Lyu, M.R.: Fused matrix factorization with geographical and social influence in location-based social networks. In: Proceedings of the Twenty-Sixth AAAI Conference on Artificial Intelligence, pp. 17–23. AAAI Press (2012)
2. Cheng, C., Yang, H., Lyu, M.R., King, I.: Where you like to go next: successive point-of-interest recommendation. In: Proceedings of the Twenty-Third international joint conference on Artificial Intelligence, pp. 2605–2611. AAAI Press (2013)
3. Cho, E., Myers, S.A., Leskovec, J.: Friendship and mobility: user movement in location-based social networks. In: Proceedings of the 17th ACM SIGKDD International Conference on Knowledge Discovery and Data Mining, pp. 1082–1090. ACM (2011)
4. Feng, S., Li, X., Zeng, Y., Cong, G., Chee, Y.M., Yuan, Q.: Personalized ranking metric embedding for next new POI recommendation. In: Proceedings of the 24th International Conference on Artificial Intelligence, pp. 2069–2075. AAAI Press (2015)
5. Gao, H., Tang, J., Hu, X., Liu, H.: Exploring temporal effects for location recommendation on location-based social networks. In: Proceedings of the 7th ACM conference on Recommender systems, pp. 93–100. ACM (2013)
6. Gao, H., Tang, J., Hu, X., Liu, H.: Content-aware point-of-interest recommendation on location-based social networks. In: Proceedings of the Twenty-Ninth AAAI Conference on Artificial Intelligence, pp. 1721–1727. AAAI Press (2015)

7. Gao, H., Tang, J., Liu, H.: gSCorr: modeling geo-social correlations for new check-ins on location-based social networks. In: Proceedings of the 21st ACM International Conference on Information and Knowledge Management, pp. 1582–1586. ACM (2012)

8. He, J., Li, X., Liao, L., Song, D., Cheung, W.K.: Inferring a personalized next point-of-interest recommendation model with latent behavior patterns. In: Thirtieth AAAI Conference on Artificial Intelligence, pp. 137–143 (2016)

9. Hu, Y., Koren, Y., Volinsky, C.: Collaborative filtering for implicit feedback datasets. In: 2008 Eighth IEEE International Conference on Data Mining, pp. 263–272. IEEE (2008)

10. Le, Q.V., Mikolov, T.: Distributed representations of sentences and documents. In: Proceedings of the 31th International Conference on Machine Learning, vol. 14, pp. 1188–1196 (2014)

11. Li, H., Hong, R., Zhu, S., Ge, Y.: Point-of-interest recommender systems: a separate-space perspective. In: 2015 IEEE International Conference on Data Mining (ICDM), pp. 231–240. IEEE (2015)

12. Li, X., Cong, G., Li, X.L., Pham, T.A.N., Krishnaswamy, S.: Rank-GeoFM: a ranking based geographical factorization method for point-of-interest recommendation. In: Proceedings of the 38th International ACM SIGIR Conference on Research and Development in Information Retrieval, pp. 433–442. ACM (2015)

13. Lian, D., Zhao, C., Xie, X., Sun, G., Chen, E., Rui, Y.: GeoMF: joint geographical modeling and matrix factorization for point-of-interest recommendation. In: ACM SIGKDD International Conference on Knowledge Discovery and Data Mining, pp. 831–840. ACM (2014)

14. Liu, P., Qiu, X., Huang, X.: Learning context-sensitive word embeddings with neural tensor skip-gram model. In: Proceedings of the 25th International Joint Conference on Artificial Intelligence. AAAI Press (2015)

15. Liu, Q., Wu, S., Wang, L., Tan, T.: Predicting the next location: a recurrent model with spatial and temporal contexts. In: Thirtieth AAAI Conference on Artificial Intelligence, pp. 194–200 (2016)

16. Liu, X., Liu, Y., Aberer, K., Miao, C.: Personalized point-of-interest recommendation by mining users' preference transition. In: Proceedings of the 22nd ACM International Conference on Information and Knowledge Management, pp. 733–738. ACM (2013)

17. Liu, X., Liu, Y., Li, X.: Exploring the context of locations for personalized location recommendations. In: Proceedings of the Twenty-Fifth International Joint Conference on Artificial Intelligence, pp. 1188–1194. AAAI Press (2016)

18. Liu, Y., Liu, Z., Chua, T.S., Sun, M.: Topical word embeddings. In: Proceedings of the 29th AAAI Conference on Artificial Intelligence. AAAI Press (2015)

19. Liu, Y., Wei, W., Sun, A., Miao, C.: Exploiting geographical neighborhood characteristics for location recommendation. In: ACM International Conference on Conference on Information and Knowledge Management, pp. 739–748. ACM (2014)

20. Mikolov, T., Le, Q.V., Sutskever, I.: Exploiting similarities among languages for machine translation (2013). arXiv:1309.4168

21. Mikolov, T., Sutskever, I., Chen, K., Corrado, G.S., Dean, J.: Distributed representations of words and phrases and their compositionality. In: Advances in Neural Information Processing Systems, pp. 3111–3119 (2013)

22. Mikolov, T., Yih, W.t., Zweig, G.: Linguistic regularities in continuous space word representations. In: Proceedings of the 2013 Conference of the North American Chapter of the Association for Computational Linguistics: Human Language Technologies, pp. 746–751 (2013)

23. Pan, R., Zhou, Y., Cao, B., Liu, N.N., Lukose, R., Scholz, M., Yang, Q.: One-class collaborative filtering. In: Proceedings of the 2008 Eighth IEEE International Conference on Data Mining, pp. 502–511. IEEE Computer Society (2008)

24. Perozzi, B., Al-Rfou, R., Skiena, S.: Deepwalk: Online learning of social representations. In: Proceedings of the 20th ACM SIGKDD International Conference on Knowledge Discovery and Data Mining, pp. 701–710. ACM (2014)

25. Recht, B., Re, C., Wright, S., Niu, F.: Hogwild: A lock-free approach to parallelizing stochastic gradient descent. In: Advances in Neural Information Processing Systems, pp. 693–701 (2011)

26. Rendle, S., Freudenthaler, C., Gantner, Z., Schmidt-Thieme, L.: BPR: Bayesian personalized ranking from implicit feedback. In: Proceedings of the Twenty-fifth Conference on Uncertainty in Artificial Intelligence, pp. 452–461. AUAI Press (2009)
27. Tang, D., Qin, B., Liu, T.: Learning semantic representations of users and products for document level sentiment classification. In: Proceedings of the 53rd Annual Meeting of the Association for Computational Linguistics and the 7th International Joint Conference on Natural Language Processing (Volume 1: Long Papers), vol. 1, pp. 1014–1023 (2015)
28. Tang, D., Qin, B., Liu, T., Yang, Y.: User modeling with neural network for review rating prediction. In: Proceedings of the 24th International Conference on Artificial Intelligence, pp. 1340–1346. AAAI Press (2015)
29. Xie, M., Yin, H., Wang, H., Xu, F., Chen, W., Wang, S.: Learning graph-based poi embedding for location-based recommendation. In: Proceedings of the 25th ACM International on Conference on Information and Knowledge Management, pp. 15–24. ACM (2016)
30. Ye, J., Zhu, Z., Cheng, H.: What's your next move: user activity prediction in location-based social networks. In: Proceedings of the 2013 SIAM International Conference on Data Mining, pp. 171–179. SIAM (2013)
31. Ye, M., Yin, P., Lee, W.C., Lee, D.L.: Exploiting geographical influence for collaborative point-of-interest recommendation. In: Proceedings of the 34th International ACM SIGIR Conference on Research and Development in Information Retrieval, pp. 325–334. ACM (2011)
32. Yin, H., Sun, Y., Cui, B., Hu, Z., Chen, L.: Lcars: A location-content-aware recommender system. In: ACM SIGKDD International Conference on Knowledge Discovery and Data Mining, pp. 221–229. ACM (2013)
33. Yuan, Q., Cong, G., Ma, Z., Sun, A., Thalmann, N.M.: Time-aware point-of-interest recommendation. In: Proceedings of the 36th international ACM SIGIR Conference on Research and Development in Information Retrieval, pp. 363–372. ACM (2013)
34. Zhang, J.D., Chow, C.Y.: GeoSoCa: exploiting geographical, social and categorical correlations for point-of-interest recommendations. In: Proceedings of the 38th International ACM SIGIR Conference on Research and Development in Information Retrieval, pp. 443–452. ACM (2015)
35. Zhang, J.D., Chow, C.Y., Li, Y.: LORE: exploiting sequential influence for location recommendations. In: Proceedings of the 22nd ACM SIGSPATIAL International Conference on Advances in Geographic Information Systems, pp. 103–112. ACM (2014)
36. Zhang, J.D., Chow, C.Y., Zheng, Y.: ORec: an opinion-based point-of-interest recommendation framework. In: Proceedings of the 24th ACM International on Conference on Information and Knowledge Management, pp. 1641–1650. ACM (2015)
37. Zhao, S., King, I., Lyu, M.R.: Capturing geographical influence in POI recommendations. In: International Conference on Neural Information Processing, pp. 530–537. Springer (2013)
38. Zhao, S., Lyu, M.R., King, I.: Aggregated temporal tensor factorization model for point-of-interest recommendation. In: International Conference on Neural Information Processing, pp. 450–458. Springer (2016)
39. Zhao, S., Zhao, T., King, I., Lyu, M.R.: Geo-Teaser: Geo-Temporal sequential embedding rank for point-of-interest recommendation. In: Proceedings of the 26th International Conference on World Wide Web Companion, pp. 153–162. International World Wide Web Conferences Steering Committee (2017)
40. Zhao, S., Zhao, T., Yang, H., Lyu, M.R., King, I.: STELLAR: spatial-temporal latent ranking for successive point-of-interest recommendation. In: Thirtieth AAAI Conference on Artificial Intelligence, pp. 315–322 (2016)

Chapter 5
STELLAR: Spatial-Temporal Latent Ranking Model for Successive POI Recommendation

Abstract Successive POI recommendation in LBSNs becomes a significant task since it helps users to navigate a large number of candidate POIs and provide the best POI recommendations based on users' most recent check-in knowledge. However, all existing methods for successive POI recommendation only focus on modeling the correlation between POIs based on users' check-in sequences, but ignore an important fact that successive POI recommendation is a time-subtle recommendation task. In fact, even with the same previous check-in information, users would prefer different successive POIs at different time. To capture the impact of time on successive POI recommendation, this chapter proposes a spatial-**te**mporal **la**tent **r**anking (STELLAR) method to explicitly model the interactions among user, POI, and time. In particular, the proposed STELLAR model is built upon a ranking-based pairwise tensor factorization framework with a fine-grained modeling of user-POI, POI-time, and POI-POI interactions for successive POI recommendation. Evaluations on two real-world datasets demonstrate that the STELLAR model outperforms state-of-the-art successive POI recommendation model about 20% in Precision@5 and Recall@5.

Keywords Successive POI recommendation · Geographical influence
Temporal influence · Latent ranking

5.1 Introduction

LBSNs such as Foursquare, Gowalla, Facebook Place, and GeoLife, become increasingly popular and provide users a new way to share their locations and experience about POIs via check-in behaviors. To help users navigate a huge number of POIs and suggest the most suitable POIs to meet their personal preferences, POI recommendation methods are developed and play an important role in LBSN services. POI recommendation learns users' preferences based on user check-in records and then predicts users' preferred POIs for recommendation. To this end, a bunch of methods has been proposed for POI recommendation recently [2, 6, 15, 28, 29].

© The Author(s), under exclusive license to Springer Nature Singapore Pte Ltd., 79
part of Springer Nature 2018
S. Zhao et al., *Point-of-Interest Recommendation in Location-Based Social Networks*,
SpringerBriefs in Computer Science, https://doi.org/10.1007/978-981-13-1349-3_5

Successive POI recommendation, as a natural extension of general POI recommendation, is proposed and has attracted great research interest recently. Different from general POI recommendation that focuses only on estimating users' preferences on POIs, successive POI recommendation provides satisfied recommendations promptly based on users' most recent checked-in location, which requires not only the preference modeling from users but also the accurate correlation analysis between POIs. Cheng et al. [3] first propose the problem of successive POI recommendation and utilize a personalized Markov chain and region localization to solve the problem. In addition, Feng et al. [5] propose a personalized metric embedding method to model the check-in sequences. However, all previous methods ignore to investigate the impact of time on successive POI recommendation.

Successive POI recommendation is a time-subtle recommendation task since at different time users would prefer different successive POIs. It is easy to imagine that a user may go to a restaurant after leaving from office at noon, while the user may be more likely to go to a gym when the user leaves office at night. However, previous successive POI recommendation methods only highlight the modeling of correlations between POIs within users' check-in sequences, but neglect to model such a time-sensitive property.

In this chapter, we try to understand the underlying mechanism of how time influences successive POI recommendation performance. To motivate this work, we first conduct an empirical analysis on two real-world LBSN datasets to verify that time is an important factor to affect users' successive POI check-in behaviors. Based on the analysis, we propose the STELLAR model to recommend a user most possible successive POIs based on the most recent check-in and the querying timestamp. The proposed STELLAR model is built upon a ranking-based pairwise tensor factorization framework with a fine-grained modeling of user-POI, POI-time, and POI-POI interactions for successive POI recommendation. To overcome the weaknesses of prior latent ranking models [1, 22, 23] that suffer from coupled interaction on POI feature, we represent each POI by three different latent feature vectors and model the three kinds of interactions separately. Moreover, the proposed STELLAR method contains two specific characteristics making it more suitable for successive POI recommendation: (1) we design a three-slice time indexing scheme to capture the temporal features of check-in behavior–preference variance and periodicity; (2) we introduce an interval-aware weight utility function to differentiate the correlations of successive check-ins, which breaks the time interval constraint in prior work [2].

To sum up, this chapter bases on the published work [38] with the following contributions.

- We propose a time-aware successive POI recommendation method–the STELLAR model, by considering the time information. In this model, we employ a new POI latent feature representation means to resolve the problem of coupled interaction. Experimental results demonstrate our STELLAR model outperforms state-of-the-art successive POI recommendation method.
- We design a three-slice time indexing scheme to represent the timestamps, which captures the user check-ins specific characteristics: preference variance and

periodicity. Experimental results show that our model better captures the temporal effect than state-of-the-art temporal models for POI recommendation.

- We introduce a new interval-aware weight utility function to differentiate successive check-ins' correlations, which improves the successive POI recommendation accuracy.

5.2 Related Work

In this section, we first review the literature of latent ranking model. Then, we show the progress of POI recommendation. Finally, we present the connection of our proposed STELLAR model and the prior work.

Latent Ranking Model. Latent ranking model is a popular solution for recommendation tasks and ranking tasks [19, 25]. In a recommendation task, latent ranking model represents user and item feature into latent vectors, and find their relations in latent subspace. In particular, Singular Value Decomposition (SVD) [19] and Nonnegative Matrix Factorization (NMF) [13] are two standard methods that exploit the latent ranking model for collaborative filtering task. Recently, [26] proposes the latent collaborative retrieval (LCR) model, which combines the retrieval and recommendation task, leading the direction of recommendations sensitive to some query condition.

POI Recommendation. POI recommendation is an important task in LBSNs. Ye et al. firstly discuss how to use memory-based methods to recommend POIs [28, 29]. In order to improve the memory-based models, advanced techniques are then leveraged to capture more information, including social and geographical influence [24, 33, 34], temporal effect [31, 35], and sequential check-ins' influence [34, 36]. On the other hand, model-based methods are proposed for the seek of scalability, most of which base on the latent ranking techniques. [2] proposes a multi-center Gaussian model to capture user geographical influence and combines it with matrix factorization (MF) model [11] to recommend POIs. [6] proposes an MF-based model which captures the temporal effect to improve performance. [27], [9], and [7] leverage user comments to improve the POI recommendation system. [15] and [17] improve POI recommendation by incorporating geographical information in a weighted regularized matrix factorization model. Instead of estimating the user preference score on POIs, [3] and [14] establish ranking models to learn the recommender system. Other techniques for POI recommendation include generative graphical models, metric learning techniques, and graph-based method. Readers may refer the papers and references therein [5, 12, 16, 30, 32].

Connection to Prior Work. We focus on successive POI recommendation, which recommends POIs on the basis of a user's most recent check-in. [3] utilizes the latent ranking model to solve the problem, while [5] employs the metric learning. Our work is most related to [3]. However, prior work does not consider the time effect on successive POI recommendation, which motivates us to propose the STELLAR model. Moreover, we propose a three-slice time indexing scheme to represent the timestamps and introduce an interval-aware weight utility function to differentiate the correlations of successive check-ins.

5.3   Data Description and Successive Check-in Analysis

Before we introduce the proposed method, in this section, we first introduce two real-world LBSN datasets and then conduct some empirical analysis on them to explore the spatial and temporal properties of users' successive check-in behaviors.

5.3.1   Data Description

We use two check-in datasets crawled from real-world LBSNs: one is Foursquare data provided in [8] and the other is Gowalla data [37]. Both contain users' check-in history from January 1, 2011 to July 31, 2011. We filter the POIs checked-in by less than five users and then choose users who check-in more than 10 times as our samples. After the preprocessing, the datasets contain the statistical properties as shown in Table 5.1.

5.3.2   Successive Check-in Analysis

Now we conduct some empirical analysis to demonstrate the spatial and temporal properties of users' successive check-in behaviors.

Spatial and temporal analysis.
Successive check-ins demonstrate significant spatial and temporal property, shown in Fig. 5.1. Figure 5.1a, b show the complementary cumulative distribution function

Table 5.1   Statistics of datasets

	Foursquare	Gowalla
#users	10,034	3,240
#POIs	16,561	33,578
#check-ins	865,647	556,453
Avg. #check-ins each user	86.3	171.7
Avg. #POIs each user	24.6	95.4
Avg. #users each POI	14.9	9.2
Density	0.0015	0.0028

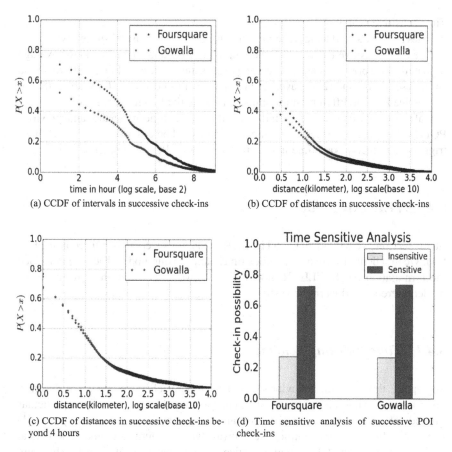

Fig. 5.1 Successive check-ins' spatial-temporal property [38]

(CCDF) of intervals and distances in successive check-ins. We verify the observation in [3] that many successive check-ins are highly correlated especially in spatial relation: over 40 and 60% successive check-in behaviors happen in less than 4 h since last check-in in Foursquare and Gowalla respectively; about 90% successive check-ins happen in less than 32 km (half an hour driving distance) in Foursquare and Gowalla. Further, we check the CCDF of distances in successive check-ins that happen beyond 4 h, shown in Fig. 5.1c. We observe although being weaker the spatial correlations still exist: about 80% successive checked-in POIs happen in less than 32 km. It is not hard to explain the phenomenon: a user always acts around his/her home or office, so the successive check-in, even independent with the last check-in, still possibly happens in the same activity area. Hence, successive checked-in POIs are spatially correlated, while successive check-ins in shorter interval contain stronger correlation.

Time sensitive analysis.

Besides the spatial and temporal contiguity, we observe that users' successive check-ins are time-sensitive behaviors. We count in all users and calculate (1) the average probability of a previous check-in leading to the same successive POI at different time (time-insensitive) and (2) the average probability of a check-in followed by different successive POIs at different timestamps (time-sensitive). Figure 5.1d shows the analytical results. We can obviously find that with different time, given the same previous POI check-in, users' successive POI check-ins would be different. This observation triggers us to incorporate time impact into successive POI recommendation.

5.4 STELLAR Model

In this section, we will detail the STELLAR model for successive POI recommendation. We first demonstrate how to index timestamps in our model. Then we introduce the formulation our STELLAR model. Finally, we demonstrate how to make the model inference and learn the system.

5.4.1 Time Indexing Scheme

To capture the check-in behavior's specific temporal characteristics, we design a novel time indexing scheme to smoothly encode a standard timestamp to a particular time id. The check-in behavior' temporal characteristics contain two aspects: (1) Periodicity [4, 31]. For example, users always visit restaurants at noon and bars at night; users check-in POIs around the office in weekdays but visit malls for shopping on weekends. (2) Preference variance [6]. Users' check-in preferences change with time. In addition, the preference variance exists in three scales: hours of a day, different days of a week, and different months of a year, which is observed in [6] but not modeled. Our proposed scheme captures the two properties in three scales as follows. First, a timestamp is divided into three slices in terms of month, weekday type, and hour slot. Next, we split a week into weekday and weekend and a day into the following four sessions: the morning session from 6:00 a.m. to 10:59 a.m., the afternoon and night session from 0:00 a.m. to 2:59 a.m. and 3:00 p.m. to 11:59 p.m., two transitive sessions that range from 3:00 a.m. to 5:59 a.m. and 11:00 a.m. to 2:59 p.m.. Further, we use 4 bits to represent the month information, 1 bit to denote weekday or weekend, and 2 bits to show the hour session. Finally, we convert the binary code into a unique decimal digit as the time ID, where the ID is in the range of 0 to 95. Figure 5.2 demonstrates the procedure of encoding an exemplary timestamp, "2011-04-05 18:10:23".

Fig. 5.2 Time encoding demonstration

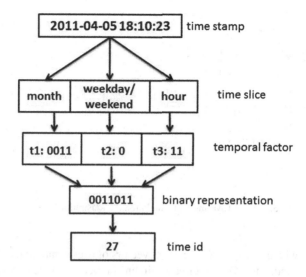

5.4.2 Model Formulation

The STELLAR system aims to provide time-aware successive POI recommendations. The task needs to learn a score function for a given user u to a candidate POI l^c at the timestamp t given his/her last check-in as a query POI l^q, which is defined as follows:

$$f(u, l^q, t, l^c),\qquad(5.1)$$

where $f : \mathcal{U} \times \mathcal{L} \times \mathcal{T} \times \mathcal{L} \to \mathbb{R}$ maps a four-tuple tensor to real values. $\mathcal{U}$, $\mathcal{L}$, and $\mathcal{T}$ denote the set of users, the set of POIs, and the set of smoothed time ids, respectively. The score value represents the "successive check-in possibility" of a user to a candidate POI at the timestamp given the query POI.

We establish a latent ranking framework to learn the score function, which employs pairwise tensor interactions to represent the following three key factors affecting users' check-in behavior: (1) the preference of a user u to a candidate POI l^c, (2) the temporal effect of time t on a candidate POI l^c, and (3) correlation of the last checked-in POI l^q and a candidate POI l^c. Correspondingly, the score value of $f(u, l^q, t, l^c)$ is determined by user-POI interaction, time-POI interaction, and POI-POI interaction together. In this case, a single vector representation for each POI is not semantically enough to capture the three different kinds of interactions. Therefore, we define a $3 \times d$ matrix to represent POI latent feature, where for each POI, there are three latent vectors used to describe the POI-user interaction, POI-time interaction and POI-POI interaction, respectively. As shown in Fig. 5.3, we formulate the function $f(u, l^q, t, l^c)$ as

$$f(u, l^q, t, l^c) = \hat{L}_{l^c,1}^T U_u + \hat{L}_{l^c,2}^T \hat{L}_{l^q,2} + \hat{L}_{l^c,3}^T T_t,\qquad(5.2)$$

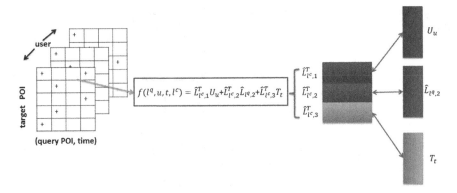

Fig. 5.3 STELLAR model formulation demonstration [38]

where U_u, $T_t \in R^d$ are latent vectors of user u and time t, $\hat{L}_{l^c,1}$, $\hat{L}_{l^c,2}$, $\hat{L}_{l^c,3} \in R^d$ are candidate POI l^c's three d-dimension vectors which correspondingly interact with users, other POIs and time labels, and $\hat{L}_{l^q,2}$ is query POI l^q's latent vector interacting to the candidate POI. To ensure the interactions are positive, all latent vectors are non-negative. Further we denote $U \in R^{d \times |\mathcal{U}|}$ as the user latent matrix and $T \in R^{d \times |\mathcal{T}|}$ as the time latent matrix. In addition, we use $\hat{L}$, a $3 \times d \times |\mathcal{L}|$ tensor, to denote the POI latent factor.

From the observations in Fig. 5.1, we find that successive POIs in a shorter interval contain a stronger correlation. To depict this observation, we introduce a weight utility function to differentiate the strong and weak correlations. The weight score value is in the range of [0,1], and the function is non-increasing with the duration of two successive check-ins. When two successive check-ins happen within a threshold interval, we assume they are highly correlated. Otherwise the correlation decreases with the increase of the time interval. In formal, we define the weight utility function as follows:

$$w = \begin{cases} 0.5 + \frac{2}{\Delta T} & \Delta T \geq s \\ 1 & otherwise \end{cases}, \tag{5.3}$$

where ΔT is the interval of successive check-ins, in unit of hour; and s is the threshold of differentiating the correlations. In our experiments, s is set as 4 to get best performance. The check-in time of query POI l^q and current time t determine the interval ΔT. So we are able to refine the score function as

$$f(u, l^q, t, l^c, w) = \hat{L}_{l^c,1}^T U_u + w \cdot \hat{L}_{l^c,2}^T \hat{L}_{l^q,2} + \hat{L}_{l^c,3}^T T_t, \tag{5.4}$$

where w is the weight value to measure the POI-POI interaction.

The STELLAR model is proposed to handle the following two challenging issues: (1) **Disastrous sparsity**. Prior methods learn a model from a tensor with only three tuples. In our formulation, we focus on tuples of four elements, (user, POI, time, POI), which increase the sparsity of the tensor significantly. (2) **Coupled interaction**. The tuples in previously proposed tensor related methods are independent. For example, in [20, 22], the tuple includes user, item, and tag, which are independent. This is easier for updating the models. However, in our constructed tensor, the tuple includes two POIs coupled in the updating. This makes the previous tensor decomposition methods [1, 22, 23] unsatisfactory. Our method represents the POI feature via a matrix and then models the three kinds of interactions separately. Further, we simplify the tensor completion problem as a combination of three low rank matrix factorization problems, which mitigates the sparsity trouble.

5.4.3 Model Inference and Learning

We make the model inference via learning the ranking order of successive check-in possibilities. Because we care more about the ranking order of the candidate POIs rather than the real values of check-in possibilities when recommending successive POIs for users. We follow the optimization criteria used in [21] and propose a pairwise ranking based objective function for the proposed STELLAR model.

We demonstrate the inference procedure following [21]. First, we suppose that the scores of $f(u, l^q, t, l^c)$ at checked-in POIs are higher than the unchecked-in counterparts. Then we define the order $l_p^c >_{u, l^q, t} l_n^c$, which means at time t, given query POI l^q, user u visits POI l_p^c but not l_n^c. Further we suffice to extract the set of all pairwise preference constraints

$$D_S := \{(u, l^q, t, l_p^c, l_n^c) | l_p^c >_{u, l^q, t} l_n^c\}. \tag{5.5}$$

Suppose the tuples in D_S are independent of each other, then to learn the parameters in the score function is to minimize the negative log likelihood of all the pair orders. Further, we add a Frobenius norm term to regularize the parameters to avoid the risk of overfitting. Then the objective function is

$$\Theta := \arg\min_{\Theta} \sum_{(u, l^q, t, l_p^c, l_n^c) \in D_S} -ln(\sigma(f(u, l^q, t, l_p^c) - f(u, l^q, t, l_n^c))) + \lambda ||\Theta||_F^2, \tag{5.6}$$

where σ is the logistic function $\sigma(x) = \frac{1}{1+e^{-x}}$, λ is the regularization parameter, and Θ denotes the parameter set, including U, T and $\hat{L}$.

We leverage the stochastic gradient decent (SGD) algorithm to learn the objective function for efficacy. Denote $\delta = 1 - \sigma(y_{u, l^q, t, l_p^c, l_n^c})$, then we get the derivative of each parameter $\theta \in \Theta$ for a tuple $(u, l^q, t, l_p^c, l_n^c)$ as

$$\frac{\partial \mathcal{O}}{\partial \theta} = \begin{cases} -\delta \cdot (\hat{L}_{l_p^c,1} - \hat{L}_{l_n^c,1}) + \lambda \cdot U_u & \theta = U_u \\ -\delta \cdot (\hat{L}_{l_p^c,3} - \hat{L}_{l_n^c,3}) + \lambda \cdot T_t & \theta = T_t \\ -\delta \cdot w \cdot (\hat{L}_{l_p^c,2} - \hat{L}_{l_n^c,2}) + \lambda \cdot \hat{L}_{l^q,2} & \theta = \hat{L}_{l^q,2} \\ -\delta \cdot U_u + \lambda \cdot \hat{L}_{l_p^c,1} & \theta = \hat{L}_{l_p^c,1} \\ -\delta \cdot w \cdot \hat{L}_{l^q,2} + \lambda \cdot \hat{L}_{l_p^c,2} & \theta = \hat{L}_{l_p^c,2} \\ -\delta \cdot T_t + \lambda \cdot \hat{L}_{l_p^c,3} & \theta = \hat{L}_{l_p^c,3} \\ \delta \cdot U_u + \lambda \cdot \hat{L}_{l_n^c,1} & \theta = \hat{L}_{l_n^c,1} \\ \delta \cdot w \cdot \hat{L}_{l^q,2} + \lambda \cdot \hat{L}_{l_n^c,2} & \theta = \hat{L}_{l_n^c,2} \\ \delta \cdot T_t + \lambda \cdot \hat{L}_{l_n^c,3} & \theta = \hat{L}_{l_n^c,3}. \end{cases} \tag{5.7}$$

To ensure the non-negativity, we project the learned parameter to non-negative value. We define the projected operator $P(\cdot) : R^d \to R^d$ as $P[x_i] = \max(0, x_i), i = 1, \ldots, d$. For each sampled tuple $(u, l^q, t, l_p^c, l_n^c) \in D_S$, we update each parameter $\theta \in \Theta$ through the derivative,

$$\theta \leftarrow P(\theta - \gamma \frac{\partial \mathcal{O}}{\partial \theta}); \tag{5.8}$$

where γ is the learning rate. To train the model, we draw the tuple from D_S via the bootstrap sampling rule, following [21]. Algorithm 5 gives the detailed procedure to learn the STELLAR model. The convergent condition is satisfied when the negative log likelihood value for a fixed sampled tuples does not decrease.

Complexity. Calculating the preference score of a tuple (u, l^q, t, l^c) costs $O(d)$, where d is the latent vector dimension. The updating procedure for each parameter is also in $O(d)$. Hence training an example (u, l^q, t, l^c) is in $O(k \cdot d)$, where k is the number of sampled unchecked POIs. Therefore, the runtime of training the model is in $O(N \cdot k \cdot d)$, where N is the number of training examples.

Algorithm 5: STELLAR model learning algorithm

Input: Training tuples $\{(u_i, l_i^q, t_i, l_i^c)\}_{i=1,\ldots,N}$
Output: $U, T, \hat{L}$
 1: Initialize $U, T, \hat{L}$
 2: **repeat**
 3:     Draw (u, l^q, t, l_p^c) uniformly from training tuples
 4:     For $s = 1, \cdots, k$, where k is #sampled unchecked POIs
 5:        Draw $(u, l^q, t, l_p^c, l_n^c)$ uniformly
 6:        Update parameters according to Eq. (3.8)
 7: **until** convergence
 8: **return** $U, T, \hat{L}$

5.5 Experiment

We conduct experiments to answer the following questions: (1) how our model performs comparing with state-of-the-art models? (2) whether our time indexing scheme works well? (3) how the parameters affect the model performance?

5.5.1 Experimental Setting

We evaluate our model on two datasets with statistics shown in Table 5.1. The system recommends a user a list of POIs, given his/her last checked-in POI and timestamp as the query. It is equivalent to solve the collaborative retrieval task [26], treating (query POI, time id, weight) as the query for each user. Following setting in [26], we extract tuples of (user, query POI, time id, weight, POI) from all successive check-ins. Here we get time id from the check-in timestamp via encoding procedure. And the weight value is calculated according to the interval between two successive check-ins through the utility function in Eq. (5.3). In order to make our model effective for future check-ins, we split the tuples into two parts, 80 and 20% according to time sequential order. So we take the first group of tuples for training and the second group for test. Finally, we measure different models through Precision@5 and Recall@5, which are general metrics for POI recommendation problem used in prior work [3, 6, 29].

5.5.2 Comparison of Methods

Our Methods. We propose three methods: **TLR**, **SLR**, and **STELLAR**. TLR and SLR methods are special cases of STELLAR, which correspondingly only ignore the POI-POI interaction and time-POI interaction.

   Baselines. We compare our proposed model with state-of-the-art latent ranking models and POI recommendation methods. Prior work [15, 17] indicates that treating the check-ins as implicit feedback is better to recommend POIs. Hence we introduce two comparative latent ranking methods that model the check-ins as

Table 5.2 Performance comparison [38]

		BPRMF	WRMF	LRT	FPMC−LR	TLAR	SLAR	STELLAR
Gowalla	P@5	0.025	0.031	0.033	0.048	0.053	0.050	**0.059**
	R@5	0.020	0.022	0.030	0.167	0.204	0.197	**0.226**
Foursquare	P@5	0.031	0.033	0.061	0.109	0.119	0.114	**0.129**
	R@5	0.027	0.028	0.053	0.347	0.373	0.368	**0.425**

implicit feedback: **WRMF** [10, 18] and **BPRMF** [21]. In addition, we introduce two state-of-the-art POI recommendation methods: **LRT** [6] and **FPMC-LR** [3]. LRT is state-of-the-art model that incorporates temporal information in a latent ranking model to improve POI recommendation. FPMC-LR is the state-of-the-art successive POI recommendation model.

5.5.3   Experimental Results

In the following, we demonstrate the performance comparison. We set latent dimension as 40, and train different models to get their best performances at appropriate parameters.

Baselines versus Our Methods. Table 5.2 shows the experimental results on Foursquare and Gowalla data. We see that: (1) Our proposed model outperforms state-of-the-art latent ranking methods and POI recommendation models. Compared with state-of-the-art successive POI recommendation method, STELLAR model gains about 22.9 and 35.3% improvement for Gowalla, and 18.3 and 22.5% improvement for Foursquare on Precision@5 and Recall@5. We observe that all models perform much better on Foursquare dataset than Gowalla dataset, even though it is sparser. The reason lies in Foursquare data contain much less POIs. (2) Our proposed models and FPMC-LR perform much better than other models, especially at recall measure. The reason lies in that these models leverage more conditions for each query. Our models recommend a user POIs given a user's recent check-in, the specific timestamp, or both; and FPMC-LR recommends POIs given a user's recent checked-in POIs. On the contrary, other three models give general recommendations.

LRT versus TLAR. The experimental results show that TLAR outperforms LRT model. Our model depicts the temporal effect with a latent feature, which gets rid of sparsity problem suffering in LRT model. Furthermore, since TLAR is a special case of STELLAR, it means that STELLAR model captures the temporal effect well from the timestamps.

FPMC-LR versus SLAR. The experimental results show that SLAR outperforms FPMC-LR model. It means SLAR model improves the recommendation performance by differentiating the correlations of successive check-ins.

5.5.4   Discussion of Time Indexing Scheme

Our three-slice time indexing scheme effectively captures the temporal effect in three scales. In order to demonstrate its efficacy, we ignore one slice to index the time and then compare their results with our model, shown in Table 5.3. 'M', 'W', and 'D' represent month, week, and day slice respectively. Our model demonstrates the best performance.

Table 5.3 Comparison of different time schemes [38]

		M+W	M+D	W+D	M+W+D
Gowalla	P@5	0.051	0.053	0.054	**0.059**
	R@5	0.207	0.208	0.219	**0.226**
Foursquare	P@5	0.118	0.120	0.121	**0.129**
	R@5	0.371	0.389	0.398	**0.425**

5.5.5 Parameter Effect

The regularization and latent dimension are important parameters to learn a latent ranking model. Figures 5.4 and 5.5 demonstrate the effect of the parameters on model performance. For simplicity, we set the same value for all latent vectors' regulariza-

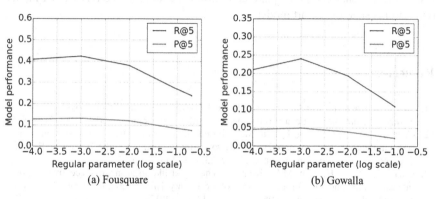

Fig. 5.4 The effect of regularization [38]

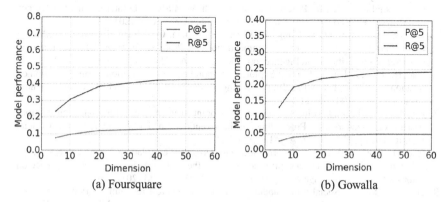

Fig. 5.5 The effect of latent dimension [38]

tions in the model. The model has best performance when $\lambda = 0.001$. The performance of Stellar steadily rises with the increase of latent vector dimension. For the trade-off of performance and computation cost, we suggest to set dimension $d = 40$.

5.6 Conclusion

In this chapter, we study the problem of successive POI recommendation. Compared with previous work, we show that successive POI recommendation is a time-subtle recommendation task. To capture the time impact, we first design a time indexing scheme to smoothly encode timestamps to particular time ids and then incorporate the time ids into our proposed STELLAR model. The STELLAR model is built upon a ranking-based pairwise interaction tensor factorization framework with a fine-grained modeling of the interactions among time, user, and POI. Experimental results on two datasets, Foursquare and Gowalla, show that the STELLAR model outperforms state-of-the-art models.

References

1. Carroll, J.D., Chang, J.J.: Analysis of individual differences in multidimensional scaling via an N-way generalization of Eckart-Young decomposition. Psychometrika **35**(3), 283–319 (1970)
2. Cheng, C., Yang, H., King, I., Lyu, M.R.: Fused matrix factorization with geographical and social influence in location-based social networks. In: Proceedings of the Twenty-Sixth AAAI Conference on Artificial Intelligence, pp. 17–23. AAAI Press (2012)
3. Cheng, C., Yang, H., Lyu, M.R., King, I.: Where you like to go next: successive point-of-interest recommendation. In: Proceedings of the Twenty-Third International Joint Conference on Artificial Intelligence, pp. 2605–2611. AAAI Press (2013)
4. Cho, E., Myers, S.A., Leskovec, J.: Friendship and mobility: user movement in location-based social networks. In: Proceedings of the 17th ACM SIGKDD International Conference on Knowledge Discovery and Data Mining, pp. 1082–1090. ACM (2011)
5. Feng, S., Li, X., Zeng, Y., Cong, G., Chee, Y.M., Yuan, Q.: Personalized Ranking Metric Embedding for Next New POI Recommendation. In: Proceedings of the 24th International Conference on Artificial Intelligence, pp. 2069–2075. AAAI Press (2015)
6. Gao, H., Tang, J., Hu, X., Liu, H.: Exploring temporal effects for location recommendation on location-based social networks. In: Proceedings of the 7th ACM conference on Recommender systems, pp. 93–100. ACM (2013)
7. Gao, H., Tang, J., Hu, X., Liu, H.: Content-aware point-of-interest recommendation on location-based social networks. In: Proceedings of the Twenty-Ninth AAAI Conference on Artificial Intelligence, pp. 1721–1727. AAAI Press (2015)
8. Gao, H., Tang, J., Liu, H.: gSCorr: modeling geo-social correlations for new check-ins on location-based social networks. In: Proceedings of the 21st ACM international conference on Information and knowledge management, pp. 1582–1586. ACM (2012)
9. Hu, B., Ester, M.: Social topic modeling for point-of-interest recommendation in location-based social networks. In: 2014 IEEE International Conference on Data Mining, pp. 845–850. IEEE (2014)
10. Hu, Y., Koren, Y., Volinsky, C.: Collaborative filtering for implicit feedback datasets. In: 2008 Eighth IEEE International Conference on Data Mining, pp. 263–272. Ieee (2008)

11. Koren, Y., Bell, R., Volinsky, C.: Matrix factorization techniques for recommender systems. Computer **42**(8), 30–37 (2009)
12. Kurashima, T., Iwata, T., Hoshide, T., Takaya, N., Fujimura, K.: Geo topic model: joint modeling of user's activity area and interests for location recommendation. In: Proceedings of the sixth ACM international conference on Web search and data mining, pp. 375–384. ACM (2013)
13. Lee, D.D., Seung, H.S.: Algorithms for non-negative matrix factorization. In: Advances in Neural Information Processing Systems, vol. 13, pp. 556–562. MIT Press (2001)
14. Li, X., Cong, G., Li, X.L., Pham, T.A.N., Krishnaswamy, S.: Rank-GeoFM: A Ranking based Geographical Factorization Method for Point-of-interest Recommendation. In: Proceedings of the 38th International ACM SIGIR Conference on Research and Development in Information Retrieval, pp. 433–442. ACM (2015)
15. Lian, D., Zhao, C., Xie, X., Sun, G., Chen, E., Rui, Y.: GeoMF: Joint geographical modeling and matrix factorization for point-of-interest recommendation. In: ACM SIGKDD International Conference on Knowledge Discovery and Data Mining, pp. 831–840. ACM (2014)
16. Liu, B., Fu, Y., Yao, Z., Xiong, H.: Learning geographical preferences for point-of-interest recommendation. In: Proceedings of the 19th ACM SIGKDD International Conference on Knowledge Discovery and Data Mining, pp. 1043–1051. ACM (2013)
17. Liu, Y., Wei, W., Sun, A., Miao, C.: Exploiting geographical neighborhood characteristics for location recommendation. In: ACM International Conference on Conference on Information and Knowledge Management, pp. 739–748. ACM (2014)
18. Pan, R., Zhou, Y., Cao, B., Liu, N.N., Lukose, R., Scholz, M., Yang, Q.: One-class collaborative filtering. In: Proceedings of the 2008 Eighth IEEE International Conference on Data Mining, pp. 502–511. IEEE Computer Society (2008)
19. Paterek, A.: Improving regularized singular value decomposition for collaborative filtering. Proc. KDD Cup Workshop **2007**, 5–8 (2007)
20. Rendle, S., Balby Marinho, L., Nanopoulos, A., Schmidt-Thieme, L.: Learning optimal ranking with tensor factorization for tag recommendation. In: Proceedings of the 15th ACM SIGKDD International Conference on Knowledge Discovery and Data Mining, pp. 727–736. ACM (2009)
21. Rendle, S., Freudenthaler, C., Gantner, Z., Schmidt-Thieme, L.: BPR: Bayesian personalized ranking from implicit feedback. In: Proceedings of the Twenty-Fifth Conference on Uncertainty in Artificial Intelligence, pp. 452–461. AUAI Press (2009)
22. Rendle, S., Schmidt-Thieme, L.: Pairwise interaction tensor factorization for personalized tag recommendation. In: Proceedings of the third ACM International Conference on Web Search and Data Mining, pp. 81–90. ACM (2010)
23. Tucker, L.R.: Some mathematical notes on three-mode factor analysis. Psychometrika **31**(3), 279–311 (1966)
24. Wang, H., Terrovitis, M., Mamoulis, N.: Location recommendation in location-based social networks using user check-in data. In: Proceedings of the 21st ACM SIGSPATIAL International Conference on Advances in Geographic Information Systems, pp. 374–383. ACM (2013)
25. Weston, J., Bengio, S., Usunier, N.: Large scale image annotation: learning to rank with joint word-image embeddings. Mach. Learn. **81**(1), 21–35 (2010)
26. Weston, J., Wang, C., Weiss, R., Berenzeig, A.: Latent collaborative retrieval. In: Proceedings of the 29th International Coference on International Conference on Machine Learning, pp. 443–450. Omni Press (2012)
27. Yang, D., Zhang, D., Yu, Z., Wang, Z.: A sentiment-enhanced personalized location recommendation system. In: Proceedings of the 24th ACM Conference on Hypertext and Social Media, pp. 119–128. ACM (2013)
28. Ye, M., Yin, P., Lee, W.C.: Location recommendation for location-based social networks. In: Proceedings of the 18th SIGSPATIAL International Conference on Advances in Geographic Information Systems, pp. 458–461. ACM (2010)
29. Ye, M., Yin, P., Lee, W.C., Lee, D.L.: Exploiting geographical influence for collaborative point-of-interest recommendation. In: Proceedings of the 34th international ACM SIGIR Conference on Research and Development in Information Retrieval, pp. 325–334. ACM (2011)

30. Yin, H., Sun, Y., Cui, B., Hu, Z., Chen, L.: Lcars: A location-content-aware recommender system. In: ACM SIGKDD International Conference on Knowledge Discovery and Data Mining, pp. 221–229. ACM (2013)
31. Yuan, Q., Cong, G., Ma, Z., Sun, A., Thalmann, N.M.: Time-aware point-of-interest recommendation. In: Proceedings of the 36th International ACM SIGIR Conference on Research and Development in Information Retrieval, pp. 363–372. ACM (2013)
32. Yuan, Q., Cong, G., Sun, A.: Graph-based point-of-interest recommendation with geographical and temporal influences. In: Proceedings of the 23rd ACM International Conference on Conference on Information and Knowledge Management, pp. 659–668. ACM (2014)
33. Zhang, J.D., Chow, C.Y.: iGSLR: personalized geo-social location recommendation: a kernel density estimation approach. In: Proceedings of the 21st ACM SIGSPATIAL International Conference on Advances in Geographic Information Systems, pp. 334–343. ACM (2013)
34. Zhang, J.D., Chow, C.Y.: GeoSoCa: Exploiting Geographical, Social and Categorical Correlations for Point-of-interest Recommendations. In: Proceedings of the 38th International ACM SIGIR Conference on Research and Development in Information Retrieval, pp. 443–452. ACM (2015)
35. Zhang, J.D., Chow, C.Y.: Ticrec: a probabilistic framework to utilize temporal influence correlations for time-aware location recommendations. IEEE Trans. Serv. Comput. 9(4), 633–646 (2016)
36. Zhang, J.D., Chow, C.Y., Li, Y.: LORE: exploiting sequential influence for location recommendations. In: Proceedings of the 22nd ACM SIGSPATIAL International Conference on Advances in Geographic Information Systems, pp. 103–112. ACM (2014)
37. Zhao, S., King, I., Lyu, M.R.: Capturing geographical influence in poi recommendations. In: International Conference on Neural Information Processing, pp. 530–537. Springer (2013)
38. Zhao, S., Zhao, T., Yang, H., Lyu, M.R., King, I.: STELLAR: Spatial-Temporal Latent Ranking for Successive Point-of-Interest Recommendation. In: Thirtieth AAAI Conference on Artificial Intelligence, pp. 315–322 (2016)

Chapter 6
Conclusion and Future Work

Abstract This chapter summarizes the main contributions of this monograph and provides several interesting future directions.

Keywords Ranking method · Online recommendation · Deep learning

6.1 Conclusion

POI recommendation is an important application in LBSNs. Due to its special geographical and temporal characteristics, POI recommendation is more challenging than traditional recommendation tasks. In order to understand the user check-in activity in LBSNs, we analyze the user mobility from geographical and temporal perspective respectively and show how to improve the POI recommendation through the geographical influence and temporal influence. Moreover, we propose two POI recommendation systems: Geo-Teaser and STELLAR.

In particular, in Chap. 2, we understand the human mobility in LBSNs from the geographical perspective and attempt to model the geographical influence for POI recommendation. In particular, we propose two models—GMM and GA-GMM to capture geographical influence. More specifically, we exploit GMM to automatically learn users' activity centers; further, we utilize GA-GMM to improve GMM by eliminating outliers. Experimental results on a real-world LBSN dataset show that GMM beats several popular geographical capturing models regarding POI recommendation, while GA-GMM excludes the effect of outliers and enhances GMM.

In Chap. 3, we study the human mobility in LBSNs from the temporal perspective. We summarize the temporal characteristics of user mobility in LBSNs in three aspects: periodicity, consecutiveness, and non-uniformness. Moreover, we observe that the temporal characteristics exist at different time scales, which cannot be modeled in prior work. To this end, we propose the ATTF model for POI recommendation to capture the three temporal features together, as well as at different time scales. Experiments on two real-world datasets show that the ATTF model achieves better performance than the state-of-the-art temporal models for POI recommendation.

S. Zhao et al., *Point-of-Interest Recommendation in Location-Based Social Networks*,
SpringerBriefs in Computer Science, https://doi.org/10.1007/978-981-13-1349-3_6

In Chap. 4, we propose the Geo-Teaser system for POI recommendation. In particular, inspired by the success of the word2vec framework to model the sequential contexts, we first propose a *temporal POI embedding* model to learn POI representations under some particular temporal state. The temporal POI embedding model captures the contextual check-in information in sequences and the various temporal characteristics on different days as well. Furthermore, we propose a new way to incorporate the geographical influence into the pairwise preference ranking method through discriminating the unvisited POIs according to geographical information. Then we develop a geographically hierarchical pairwise preference ranking model. Finally, we propose a unified framework to recommend POIs combining these two models. To verify the effectiveness of our proposed method, we conduct experiments on two real-life datasets. Experimental results show that the Geo-Teaser model outperforms state-of-the-art models.

In Chap. 5, we propose the STELLAR model for time-aware successive POI recommendation. In particular, the proposed STELLAR model is built upon a ranking-based pairwise tensor factorization framework with a fine-grained modeling of user-POI, POI-time, and POI-POI interactions for successive POI recommendation. In addition, we design a novel three-slice indexing scheme to represent the timestamps, which captures the user check-ins' specific characteristics: preference variance and periodicity. Moreover, we propose a new interval-aware weight utility function to differentiate successive check-ins' correlations, which breaks the time interval constraint in prior work. Evaluations on two real-world datasets demonstrate that the STELLAR model outperforms state-of-the-art successive POI recommendation model about 20% in Precision@5 and Recall@5.

6.2 Future Work

A bunch of studies has been proposed for POI recommendation. Summarizing the existing work, we point out the trends and new directions in three possible aspects: ranking-based model, online recommendation, and deep learning based recommendation.

6.2.1 Ranking-Based Model

Several ranking-based models [7, 13, 25] have been proposed for POI recommendation recently. Most of the previous methods attempt to estimate the user check-in probability over POIs [4, 8, 9]. However, for the POI recommendation task, we do not care about the predicted check-in possibility value but the preference order. Some work has proved that it is better for the recommendation task to learn the order rather than the real value [12, 19–21]. BPR loss [19] and WARP loss [20, 21] are two popular pairwise loss criteria to learn the ranking order. Researchers in [5, 7, 25] leverage the BPR loss to learn a POI recommendation model, and Li et al. [13] use the WARP loss. Also, He et al. [10] propose a list-wise ranking model for POI recommendations. The existing work using ranking-based model has shown its advantage in model performance. Then, learning to rank, as an important technique for information retrieval [3, 15], may be used more for POI recommendation to improve performance in the future.

6.2.2 Online Recommendation

The online POI recommendation model has advantages over offline models in two aspects: cold-start problem and adaptability to the user behavior variance. Most of the previous work recommends POIs via the offline model. Hence, the previous work is apt to suffer the two problems: (1) cold-start problem, the proposed model performs not satisfying for new users or users who have only a few check-ins; (2) user behavior variance, the proposed model, may perform awfully if a user's behavior changes since it learns user behavior according to historical records. Researchers in [1, 23] utilize offline model and online recommendation to improve the recommendation results. However, there is no work using online model for POI recommendation. In fact, online recommendation models based on multi-armed bandits [2] have been proposed for movie recommendation and advertisement recommendation [18, 26]. In the future, online recommendation methods will be a new direction for POI recommendation.

6.2.3 Deep Learning Based Recommendation

Inspired by the success of deep learning, the neural network method has been used to model the check-in sequences. Liu et al. [14] employ RNN to find the sequential correlations. In addition, several studies [6, 16, 22, 24] leverage the embedding learning for POI recommendations. Liu et al. [16] model the check-in sequences through the word2vec framework [17] to capture the sequential contexts. Xie et al. [22] propose a graph-based framework for POI recommendations to systematically model the POI, user, and time relations in an embedding space and learn the representations through the word2vec framework. Moreover, we [24] propose a temporal POI embedding based on Skip-Gram model [17] and combine it with a geographically pairwise user preference ranking model to recommend POIs. In the future, more advanced techniques, such as LSTM [11], can be used for POI recommendation.

References

1. Bao, J., Zheng, Y., Mokbel, M.F.: Location-based and preference-aware recommendation using sparse geo-social networking data. In: Proceedings of the 21st ACM SIGSPATIAL International Conference on Advances in Geographic Information Systems, pp. 199–208. ACM (2012)
2. Bubeck, S., Cesa-Bianchi, N., et al.: Regret analysis of stochastic and nonstochastic multi-armed bandit problems. Foundations and Trends ®in Machine Learning **5**(1), 1–122 (2012)
3. Cao, Z., Qin, T., Liu, T.Y., Tsai, M.F., Li, H.: Learning to rank: from pairwise approach to listwise approach. In: Proceedings of the 24th international conference on Machine learning, pp. 129–136. ACM (2007)
4. Cheng, C., Yang, H., King, I., Lyu, M.R.: Fused matrix factorization with geographical and social influence in location-based social networks. In: Proceedings of the Twenty-Sixth AAAI Conference on Artificial Intelligence, pp. 17–23. AAAI Press (2012)
5. Cheng, C., Yang, H., King, I., Lyu, M.R.: A unified point-of-interest recommendation framework in location-based social networks. ACM Trans. Intell. Syst. Technol. (TIST) **8**(1), 10 (2016)
6. Feng, S., Cong, G., An, B., Chee, Y.M.: POI2Vec: Geographical latent representation for predicting future visitors. In: Proceedings of the Thirty-First AAAI Conference on Artificial Intelligence, February 4-9, 2017, San Francisco, California, USA., pp. 102–108 (2017)
7. Feng, S., Li, X., Zeng, Y., Cong, G., Chee, Y.M., Yuan, Q.: Personalized ranking metric embedding for next new poi recommendation. In: Proceedings of the 24th International Conference on Artificial Intelligence, pp. 2069–2075. AAAI Press (2015)
8. Gao, H., Tang, J., Hu, X., Liu, H.: Exploring temporal effects for location recommendation on location-based social networks. In: Proceedings of the 7th ACM Conference on Recommender Systems, pp. 93–100. ACM (2013)
9. Gao, H., Tang, J., Hu, X., Liu, H.: Content-aware point-of-interest recommendation on location-based social networks. In: Proceedings of the Twenty-Ninth AAAI Conference on Artificial Intelligence, pp. 1721–1727. AAAI Press (2015)
10. He, J., Li, X., Liao, L.: Category-aware next point-of-interest recommendation via Listwise Bayesian personalized ranking. In: Proceedings of the 26th International Joint Conference on Artificial Intelligence, pp. 1837–1843. AAAI Press (2017)
11. Hochreiter, S., Schmidhuber, J.: Long short-term memory. Neural Comput. **9**(8), 1735–1780 (1997)
12. Lee, J., Bengio, S., Kim, S., Lebanon, G., Singer, Y.: Local collaborative ranking. In: Proceedings of the 23rd International Conference on World Wide Web, pp. 85–96. ACM (2014)
13. Li, X., Cong, G., Li, X.L., Pham, T.A.N., Krishnaswamy, S.: Rank-GeoFM: a ranking based geographical factorization method for point-of-interest recommendation. In: Proceedings of the 38th International ACM SIGIR Conference on Research and Development in Information Retrieval, pp. 433–442. ACM (2015)
14. Liu, Q., Wu, S., Wang, L., Tan, T.: Predicting the next location: a recurrent model with spatial and temporal contexts. In: Thirtieth AAAI Conference on Artificial Intelligence, pp. 194–200 (2016)
15. Liu, T.Y.: Learning to rank for information retrieval. Found. Trends Inf. Retr. **3**(3), 225–331 (2009)
16. Liu, X., Liu, Y., Li, X.: Exploring the context of locations for personalized location recommendations. In: Proceedings of the Twenty-Fifth International Joint Conference on Artificial Intelligence, pp. 1188–1194. AAAI Press (2016)
17. Mikolov, T., Sutskever, I., Chen, K., Corrado, G.S., Dean, J.: Distributed representations of words and phrases and their compositionality. In: Advances in Neural Information Processing Systems, pp. 3111–3119 (2013)
18. Qin, L., Chen, S., Zhu, X.: Contextual combinatorial bandit and its application on diversified online recommendation. In: Proceedings of the 2014 SIAM International Conference on Data Mining, pp. 461–469. SIAM (2014)

19. Rendle, S., Freudenthaler, C., Gantner, Z., Schmidt-Thieme, L.: BPR: Bayesian personalized ranking from implicit feedback. In: Proceedings of the Twenty-Fifth Conference on Uncertainty in Artificial Intelligence, pp. 452–461. AUAI Press (2009)

20. Usunier, N., Buffoni, D., Gallinari, P.: Ranking with ordered weighted pairwise classification. In: Proceedings of the 26th Annual International Conference on Machine Learning, pp. 1057–1064. ACM (2009)

21. Weston, J., Bengio, S., Usunier, N.: Large scale image annotation: learning to rank with joint word-image embeddings. Mach. Learn. **81**(1), 21–35 (2010)

22. Xie, M., Yin, H., Wang, H., Xu, F., Chen, W., Wang, S.: Learning graph-based POI embedding for location-based recommendation. In: Proceedings of the 25th ACM International on Conference on Information and Knowledge Management, pp. 15–24. ACM (2016)

23. Yin, H., Sun, Y., Cui, B., Hu, Z., Chen, L.: Lcars: a location-content-aware recommender system. In: ACM SIGKDD International Conference on Knowledge Discovery and Data Mining, pp. 221–229. ACM (2013)

24. Zhao, S., Zhao, T., King, I., Lyu, M.R.: Geo-Teaser: geo-temporal sequential embedding rank for point-of-interest recommendation. In: Proceedings of the 26th International Conference on World Wide Web Companion, pp. 153–162. International World Wide Web Conferences Steering Committee (2017)

25. Zhao, S., Zhao, T., Yang, H., Lyu, M.R., King, I.: STELLAR: spatial-temporal latent ranking for successive point-of-interest recommendation. In: Thirtieth AAAI Conference on Artificial Intelligence, pp. 315–322 (2016)

26. Zhao, T., King, I.: Constructing reliable gradient exploration for online learning to rank. In: Proceedings of the 25th ACM International on Conference on Information and Knowledge Management, pp. 1643–1652. ACM (2016)

Index

Printed in the United States
By Bookmasters